The Shaping of Greenland's Resource Spaces

The book examines ideas about the making and shaping of Greenland's society, environment, and resource spaces.

It discusses how Greenland's resources have been extracted at different points in its history, shows how acquiring knowledge of subsurface environments has been crucial for matters of securitisation, and explores how the country is being imagined as an emerging frontier with vast mineral reserves. The book delves into the history and contemporary practice of geological exploration and considers the politics and corporate activities that frame discussion about extractive industries and resource zones. It touches upon resource policies, the nature of social and environmental assessments, and permitting processes, while the environmental and social effects of extractive industries are considered, alongside an assessment of the status of current and planned resource projects. In its exploration of the nature and place of territory and the subterranean in political and economic narratives, the book shows how the making of Greenland has and continues to be bound up with the shaping of resource spaces and with ambitions to extract resources from them. Yet the book shows that plans for extractive industries remain controversial. It concludes by considering the prospects for future development and debates on conservation and Indigenous rights, with reflections on how and where Greenland is positioned in the geopolitics of environmental governance and geo-security in the Arctic.

This book will be of great interest to students and scholars of environmental anthropology, geography, resource management, extractive industries, environmental governance, international relations, geopolitics, Arctic studies, and sustainable development.

Mark Nuttall is Professor and Henry Marshall Tory Chair in the Department of Anthropology at the University of Alberta, Canada, and Adjunct Professor at Ilisimatusarfik/University of Greenland and the Greenland Climate Research Centre in Nuuk. His books include *Climate, Society and Subsurface Politics in Greenland: Under the Great Ice* (Routledge, 2017), *The Scramble for the Poles: the Geopolitics of the Arctic and Antarctic* (with Klaus Dodds, 2016), and *The Arctic: What Everyone Needs to Know* (with Klaus Dodds, 2019). He is editor of the *Encyclopedia of the Arctic* (Routledge, 2005), and co-editor of *Anthropology and Climate Change* (Routledge, 2016) and the *Routledge Handbook of the Polar Regions* (Routledge, 2018). He is a Fellow of the Royal Society of Canada.

Routledge Studies of the Extractive Industries and Sustainable Development

Oil and National Identity in the Kurdistan Region of Iraq
Conflicts at the Frontier of Petro-Capitalism
Alessandro Tinti

The Anthropology of Resource Extraction
Edited by Lorenzo D'Angelo and Robert Jan Pijpers

Andean States and the Resource Curse
Institutional Change in Extractive Economies
Edited by Gerardo Damonte and Bettina Schorr

Stakeholders, Sustainable Development Policies and the Coal Mining Industry
Perspectives from Europe and the Commonwealth of Independent States
Izabela Jonek-Kowalska, Radosław Wolniak, Oksana A. Marinina and Tatyana V. Ponomarenko

The Social Impacts of Mine Closure in South Africa
Housing Policy and Place Attachment
Lochner Marais

Local Communities and the Mining Industry
Economic Potential and Social and Environmental Responsibilities
Edited by Nicolas D. Brunet and Sheri Longboat

The Shaping of Greenland's Resource Spaces
Environment, Territory, Geo-Security
Mark Nuttall

For more information about this series, please visit: www.routledge.com/ Routledge-Studies-of-the-Extractive-Industries-and-Sustainable-Development/ book-series/REISD

The Shaping of Greenland's Resource Spaces

Environment, Territory, Geo-Security

Mark Nuttall

Routledge
Taylor & Francis Group

LONDON AND NEW YORK

earthscan
from Routledge

First published 2024
by Routledge
4 Park Square, Milton Park, Abingdon, Oxon OX14 4RN

and by Routledge
605 Third Avenue, New York, NY 10158

Routledge is an imprint of the Taylor & Francis Group, an informa business

British Library Cataloguing-in-Publication Data
A catalogue record for this book is available from the British Library

ISBN: 978-1-032-00748-9 (hbk)
ISBN: 978-1-032-00751-9 (pbk)
ISBN: 978-1-003-17542-1 (ebk)

DOI: 10.4324/9781003175421

Typeset in Goudy
by codeMantra

Contents

Figures

Introduction

Kalaallit Nunaat – Greenland – is in the crosscurrents of Arctic change. Global interest in the world's largest island, which is a self-governing territory of the Kingdom of Denmark, has intensified over the last couple of decades. Greenland has assumed sharper international visibility and geopolitical prominence. This is perhaps most markedly so in relation to global concern over environmental crisis and climate breakdown, melting ice, the presence of extractive industries, and the changing nature of the polar regions in the Anthropocene. The high latitudes of the world are warming rapidly and scientific research points increasingly and with more urgency to how circumpolar ecosystems are turbulent and approaching tipping points. The signs of a changing Arctic climate are increasingly apparent through the mass loss of glacial ice, the pronounced retreat and thinning of sea ice during winter as well as summer, thawing permafrost, coastal erosion, an intensification of extreme and stormy weather, and changes to the migration routes and population sizes of a number of animal and fish species.

Shifting weather patterns are having notable effects on northern environments, on Indigenous and local livelihoods, and on wider northern regional economies (AMAP 2012, 2017; Meredith and Summerkorn 2019). One of the most obvious areas of global attention is the continued melting of parts of Greenland's ice sheet, or inland ice, and the retreat and reduction of many of its outlet glaciers. Known as *sermersuaq* – 'the great ice' – in Kalaallisut (Greenlandic), the inland ice covers an area of 1.7 million km^2, which is 80% of the country's landmass. The melt and runoff is already influencing the world's weather patterns and contributing to global sea level rise (Box et al. 2022).

For those who do not live in the Arctic, awareness of melt and climate breakdown may more likely come from reading media reports, viewing satellite images of receding ice, or gazing with alarm at photographs of starving polar bears, but the effects of climate change are sensed, felt, and experienced in immediate, embodied ways by the people for whom their Arctic surroundings are homelands. In Northwest Greenland, for example, in the districts of Upernavik and Avanersuaq, where I have continued to do anthropological research since first living there in the late 1980s, the entire land-fast sea ice ecosystem has undergone significant transformation over the past few decades and glacial melt is altering water temperature, influencing ocean productivity, and affecting marine mammals and fish.

DOI: 10.4324/9781003175421-1

There is a large, ever-growing body of scientific literature that provides stark evidence of this (e.g. see Ballinger et al. 2022), but my friends who live in the region's coastal communities do not need scientists to tell them that their surroundings are shifting in ways that are unprecedented and worrying.

Sea ice (*siku* in Kalaallisut or *hiku* in the Inuktun spoken in the Avanersuaq area), for instance, has been and continues to be central to people's lives for several months of the year. It allows for a range of possibilities for hunting, fishing, and mobility. Ice connects communities in a significantly different way than the open water does, allowing for new travel routes and enabling a range of configurations of engagement with the environment. Community life, networks of family, kinship, and close social relatedness, patterns of sharing meat and fish, and ways of thinking about and relating to the non-human have all been bound up with and inextricably connected to an annual seasonal round of hunting and fishing activities, of extensive travel on the sea ice by dog sledge in winter and spring and by boat along the coast, and through inner fjords and around dense patterns of islands in summer and autumn (Hastrup 2009; Nuttall 1992). Yet the trend over the last twenty to twenty-five years or so – one that is both observed, studied, and experienced – has been for ice to form later in the autumn and break up earlier in the spring. The apparent stability, fixity, and fastness of sea ice throughout the winter can no longer be taken for granted, while an increasingly rougher sea during summer and autumn throws up its own challenges. Projections of future sea ice melt in Northwest Greenland point towards continued declining drift ice in Baffin Bay and the decline and continued thinning of land-fast ice. The inflow of warmer Atlantic water and the discharge of meltwater from glaciers are processes that also influence the temperature of the sea and are affecting the coastal ecosystem (Andresen et al. 2014; Muilwijk et al. 2022).

In May 2017, for example, a friend from Kangersuatsiaq (which is located to the south of Upernavik town) told me how he was fishing from his small boat for Greenland halibut and cod on 21 December 2016 just to the south of the village. It was very cold, he said, and he was fishing and moving around the area in the midwinter darkness, but there was no sea ice. Nothing had begun to form. He thought this to be unusual and concerning, and he was able to take his catch to Upernavik to sell the following day. There is no fish landing or processing facility in Kangersuatsiaq, so a journey to Upernavik in a skiff fitted with an outboard engine takes around three hours. When the ice did form a few weeks later in January it was almost gone by early March. Other people in the village described this ice as not being good ice, though. Even when it did finally form, they could not, for the most part, get out to the fjords during winter by dog sledge or snowmobile to fish as the ice was not solid enough or had not taken shape at all in parts, while hunting seals by open boat was hindered by the moving pack that was there. This is not an isolated example of what people are experiencing, and I have heard similar accounts of being able to hunt and fish on ice-free coastal waters in midwinter since then. While some people are adapting to the changing marine environment, and anticipating both its challenges and opportunities, it has also been one of the reasons why more people are deciding to leave the village and move to Upernavik, as well as to other towns including Nuuk, Greenland's growing capital (Nuttall 2022).

The marine ecosystem supports the livelihoods of many people in Northwest Greenland, as it does for communities elsewhere along the country's coasts, so thinning and diminishing ice cover and a warming sea have consequences for hunting and fishing activities, for mobility, and for local economies that are far-reaching. When hunters and fishers talk about changes in the weather and the environment, they will generally tell of their experiences of thinning, patchy ice, or brittle ice (*sikulaaq*), and areas of open water where the surface of the sea no longer freezes (*sikujuippoq*). But I have also travelled with hunters by dog sledge on the sea ice along the northwest coast during several winters and spring seasons before the climate change effects that are observable now were becoming so apparent. Some of these were long hunting trips where we spent several weeks at a time on the ice. As an anthropologist, I have come to understand what it is that people have to deal with, and I can put this in the context of decades of getting to know life in northern Greenland through long-term ethnographic research. For example, a winter or spring journey north from Upernavik to Qaanaaq across Melville Bay (Qimusseriarsuaq in Kalaallisut, which means 'the great dog sledging place'), would be undertaken to hunt polar bears or to visit relatives living in communities along the way. Since the early 2000s, it has been a difficult, if impossible, sledge route because of poor ice conditions.

Indigenous anticipatory knowledge is challenged by these transformations and the flexibility that has long been characteristic of life along the coast is reduced. Historian Joy Parr has written that bodies are archives of sensory knowledge (Parr 2010); in this way, and as I have spoken with people in several parts of Greenland about, environmental change disrupts people's embodied knowledge of the world with profound implications for identity and sense of place. This does not mean that people consider themselves to be passive victims of climate change, however. Inuit in Northwest Greenland – as indeed elsewhere along the country's coasts – have experienced a long history of social, economic, and environmental change that has brought profound disruption to their lives. While much of this has been troubling and deeply traumatic, they have adapted and persevered in the face of all of this, which perhaps provides a clue to how they contend with the effects of rapid climate change today.

Understanding climate change and how people live with the precarious reality its effects usher in is central to much of the research I do with communities in northern Greenland as well as in large towns such as Nuuk. I am interested too in the different scales of attention given to Greenland and how daily life there is caught up in and affected by more global narratives about a rapidly changing Arctic. The global gaze that has come to settle on Greenland has intensified in recent years. For one thing, climate change plays a significant role in how Greenland is represented by scientists, Indigenous leaders, politicians, environmentalists, activists, and the media as a place under threat. It is an Arctic region that matters to the world and for action concerning sustainability, climate change, and environmental protection as well as Indigenous rights, entitlements, and security in the circumpolar north.

With Arctic ecosystems in flux, the implications are not just environmental, social and economic. They are geopolitical and geostrategic as well. The emergence of what is increasingly referred to as the 'new Arctic' as a key geopolitical region under conditions of melt, climate breakdown, and sharper international interest is now a commonplace storyline for academic research, the media, and policymakers. But the trope of a melting Arctic and the argument that there is increasing accessibility to what is imagined and described a new resource frontier is central to contemporary manifestations of Arctic geopolitics, and so discussion in Greenland also focuses on how global warming brings opportunities. Hunting and fishing livelihoods may be increasingly precarious and biodiversity is threatened as the ice melts and the sea warms, but international companies are scoping out the potential for extracting minerals in new resource spaces. Tourism is also growing, and conversations about the shaping of the country's future are centred largely on the development of Nuuk as an Arctic metropolis with greater global connectivity, the greater urbanisation of other towns (along with a depopulation of small settlements), the construction of infrastructure such as new airports, health, education, housing, and other social issues. Greenland is becoming an increasingly cosmopolitan society. Yet, as Nuuk grows (planners anticipate that more than half the country's population will be living there within the next couple of decades), many villages and small settlements – like the ones I know along the northwest coast – are struggling economically. Debate in Greenland's parliament is once more focused on their viability and their very future.

Against the backdrop of Arctic change, Greenland is in the foreground of debate about resources and sustainability in the Anthropocene Arctic. It is being imagined as a space of economic possibility and resource speculation. Although international mining and oil companies have been going there for decades, many new players have been enticed in more recent years by the seductive idea of Greenland as an emerging resource frontier with vast mineral reserves. Developing a mining industry and, until recently, encouraging hydrocarbon exploration have remained stated aims of the various coalition permutations of Greenland's self-rule government. As Greenland pushes forward with economic and financial planning for how to reduce the country's great reliance on the annual block grant it receives from Denmark – and as politicians reflect on the possibilities for political independence – economic policy and decisions about infrastructure development have been shaped greatly, especially since the early 2000s, by the anticipation of a resource boom as well as by other business prospects connecting the country to wider networks of global commerce. Thinning ice and glacial retreat may mean that hunters and fishers in some parts of the country face restrictions and frustrations in the customary movements they make around their increasingly liquescent localities, but politicians and business leaders seize on the possibilities climate change has for further empowerment, nation-building, state formation, and autonomy.

This book explores the making of Greenland through an investigation of some aspects of the extraction of resources in Greenland in both historical and contemporary perspective. In doing so, it looks at the shaping of Greenland's resource spaces, which come in many forms. The living beings in the sea and on land have provided the basis for Inuit cultural survival since the first Paleo-Inuit groups moved across to Greenland from what is now the Canadian High Arctic. Inuit today continue to live and move around in surroundings that are brought into being through practices of hunting and fishing and engagement between the human and non-human. Other resource spaces have been imagined and shaped, though, that involve configurations and reconfigurations of territory and the resources that are defined, classified, given value within them, and extracted from them. They are related to the expansion of capital and are often sites in which contentious encounters of authority, power, control, and sovereignty occur (Rasmussen and Lund 2018). It is not just about getting minerals or hydrocarbons out of the ground, though. I examine how Greenland was viewed as a vast resource space for Danish trade activities from the eighteenth century and how extracting knowledge of its subsurface properties was crucial to activities that turned Greenland into a securitised space, one that extended from the depths of the sea, across and under the ice, and up into the skies above. Much of the book, however, is concerned with a discussion of how, in Greenland today, extractive terrain is being mapped and how resource spaces have been designated and marked out for oil exploration and the surveying and assessment of potential mining ventures, both on land and offshore. These activities are often written about as being spurred on by a sense of the increasing accessibility of what, until now, have been considered remote, frozen Arctic areas. Now, as sea ice melts and glaciers recede, topography and climate are no longer considered obstacles to extraction and getting resources to global markets. Indeed, melt is seen to enable new forms of mobility through and across Arctic waters and landscapes for geophysical exploration crews, seismic survey vessels, and the flows of capital, investment, expertise, skills, and labour needed for resource industry projects and the unearthing of minerals and hydrocarbons.

In mining company speak, Greenland is considered to be one of the world's last frontiers for extraction. This is a view that is reproduced with regularity by more popular and media narratives about scrambles for resources and territory in a new Arctic (e.g. Dodds and Nuttall 2016), but the reality is different. Political narratives, scientific knowledge, and economic assessments about the underground and resource wealth have long informed ideas and efforts concerned with the making and shaping of Greenland. The exploration of Greenland's geology and the assessment of the country's resource potential have a long history. Understanding the subterranean nature of Greenland for economic reasons as well as for purposes of defence and securitisation predates any contemporary awareness and understanding of climate change as global crisis and the anxieties – as well as the feelings of excitement some have – that easier access to remote areas will follow from melting ice. That said, the idea of an increasingly less frozen and more accessible Greenland, a place where riches can be unearthed and revealed as the ice melts and disappears, has long been used as a persuasive device by Greenlandic politicians as they aspire to develop an extractive industry sector, market the country's resource potential, and attract international interest and investment (Nuttall 2008, 2017).

Observing how Greenlandic politics has been more closely influenced in recent years by the presence of the extractive industry sector, I have been concerned in much of my research with understanding what follows when the self-rule government focuses on developing working relationships with foreign mining and oil companies and overseas investors. Pursuing an extractive imperative, often framed as a policy solution to achieving greater autonomy and meeting socio-economic challenges (cf. Arsel, Hogenbook and Pellegrino 2016), democratically elected Greenlandic parliaments and governments have set out strategies for how oil, gas, and mineral exploitation are intended to contribute to the overall sustainable development of Greenlandic society. Politicians, civil servants, and private businesses promote Greenland as a place that is open for business. They invest considerable energy and effort – while leaving their own considerable carbon footprint – in globetrotting to attract mining companies and encouraging them to apply for prospecting and exploration licences (Bjørst 2020; Sejersen 2020).

The campaign to promote business opportunities in Greenland works to place emphasis on marketing the country's resource potential, but it is being broadened to target many more links along the global supply chain, from exploration companies to those who purchase the materials of extraction and those who manufacture the products that require the kind of critical minerals found in Greenland's subsoil that the world needs as it transitions to a post-carbon future. And while international companies are scoping out the ground – and the seabed – for extracting minerals (and until recently, hydrocarbons), they are not the only ones eyeing Greenland's resource potential. There are fishing booms in coastal waters for shrimp and mackerel. Tourism is growing and new business enterprises are emerging. Other initiatives are testing the potential for the export of fresh water to supply regions affected by drought and are examining what possibilities may arise from the use of glacial rock flour in fertiliser products that can regenerate arid farmland in sub-Saharan Africa.

In the nineteenth century, field surveys and assessments of the potential for mining were carried out, even if extractive industry was not a primary focus for Danish economic ventures at the time, which were based on the trade of marine mammal products, fish, and furs (Nuttall 2017), while in the 1900s, Danish administrators laid down plans for the industrialisation of Greenland's natural resources focused on fisheries and mines (Priebe 2017). Commercial prospecting and exploration for minerals and hydrocarbons has been taking place since the 1960s and specific government policy on mining as critical to Greenland's economy has been enacted since the 1990s. The interesting thing about all this is that, although there have been some major mining projects in the past – such as a cryolite mine at Ivittuut in South Greenland (open from 1859 to 1987), a coal mine at Qullissat on Disko Island (from 1924 to 1972), a lead-zinc mine at Maarmorilik in Ummannaq fjord (beginning in 1938, but as a major operation from 1973 to 1990), a gold mine near Nanortalik near the country's southern tip (from 2004 to 2013), and a lead and zinc mine at Mesters Vig in Northeast Greenland (1956–1963) – Greenland is not yet a mining nation on the scale that has been envisaged, whether historically or in more contemporary times. Some projects that a decade ago seemed especially promising and possible have not been built, as I discuss in this book. There is also a sense that Greenland is an unfinished resource frontier, not just a new one (cf. Carse and Kneas 2019). Resource talk often seems to revolve about matters of speculation.

This book is not a study of the social, economic, and environmental impacts of a particular mine in Greenland, and it is concerned with more than just extractive industry. As I discuss later, only a few projects are currently underway, although many more are in the early stages of prospecting and exploration. Espig and de Rijke (2018) point out that global energy networks make demands of anthropologists to rethink the nature of their ethnographic 'field sites' and consider and understand extractive industries and their wider reach, as well as their social, economic, and environmental impacts in relation to global interconnectedness. As Hein (2018) puts it in relation to the oil industry, constellations of actors, which include corporations and nations, actively shape the geographical, physical, and material spaces that appear to be disconnected and geographically distant. In a similar vein, Schritt and Behrends (2018: 212) talk about 'oil zones' in West Africa – and the extraction that occurs in them – as trans-territorial spaces. I think of Greenland as becoming a component of the planetary mine (Arboleda 2020) in the ways that resources enter the political imagination, how spaces of extraction are being made, and how capital is invested in Greenlandic ventures. Building on Mazzan Labban's idea of the planetary mine (Labban 2014), Martín Arboleda draws on his own research on mining in Chile's Atacama Desert to inform a theoretical framework that sheds light on how geographies of extraction are 'entangled in a global apparatus of production and exchange that supersedes the premises and internal dynamics of a proverbial world system of cores and peripheries defined exclusively by national borders' (Arboleda ibid.: 5). The point here, he argues, is that a mine is not a 'discrete sociotechnical object but a dense network of territorial infrastructures and spatial technologies widely distributed across space' (ibid.).

As Arboleda does in his book *Planetary Mine*, one of my intentions with writing this book is to consider what the Greenlandic case can tell us about resource extraction by understanding the 'changing configurations of resource frontiers' and other spaces and what the global reach of industrialisation and extraction mean (ibid.: 14). The changing configuration of Greenland as a resource frontier has been a historical process from the extraction of living marine and land resources on a commercial scale, to geological mapping, to the investigation of subsurface space for reasons of both economic development and securitisation, and to the potential for mining and extracting hydrocarbons. Global demand for critical metals and raw materials, along with sharper political and public attention in Greenland focused on self-determination and state formation, bring a different kind of imperative to examining the plans, narratives, and debates that crystallise around the quest for drilling and digging into Greenland's subsurface.

Geological strata and the subterranean have become far more central to ideas about sustainability and the strengthening of Greenland's economy since the country achieved a greater degree of self-government in 2009 and acquired ownership of its subsurface resources at the beginning of 2010. The geological mapping and the visualisation and representation of the vertical and volumetric dimensions (cf. Elden 2013) as well as the economic potential of the subsurface and the seabed and the classification and assessment of their hydrocarbon and mineral assemblages are crucial elements of this policy (Dodds and Nuttall 2018; Nuttall 2017). Geology and deep time processes form the bedrock for discussions and debates concerning Greenland's future and state-making practices, but they also have implications for how we understand the effect of geopolitical realignments and securitisation on the wider Arctic.

Despite a fall in global commodity prices over the last few years as well as the effects of other global processes on plans for resource development in the Arctic (including a shift away from fossil fuels as part of a global low carbon economy agenda, and the technical challenges of Arctic resource extraction, regardless of the impression that access to resources may be easier with less ice around), interest in the region's minerals has not abated. Some companies and investors still hope there may be a resurgence in exploration for hydrocarbons in parts of the Arctic if oil and gas prices become more favourable, notwithstanding the associated challenges for governance. In July 2021, though, Greenland's government placed a moratorium on new exploratory licences for hydrocarbons and has since emphasised the importance of mining for sustainability, with a focus on critical minerals and green energy production. Planning is also underway for underground carbon capture and storage projects.

In this book, I focus on how Greenland's resources (different kinds and not just minerals) have been extracted at different points in its history and how extractive industries are said to hold promise for the country's future. This involves a discussion about the making of resources and the spaces in which they are located, environmental change, sustainability, and the geopolitics of the subsurface. In doing so, I continue my investigation of how knowledge of Greenland's subsoil and ice has been produced – and why and for whom – but also with how subterranean resource spaces and the mineral reserves that are defined, measured, and classified within them have become social, political, and ideological sites. But this history is also one of the making and designing of Greenlandic society, towns, and settlements. In the chapters that follow, I draw largely on historical and ethnographic work done over the last twelve years or so, but I also build upon my research over many decades on understanding human-environment relations and human and non-human interactions in Greenland as well as with more recent investigations on the social and cultural effects of climate change and the politics of extractive projects (e.g. Nuttall 1992, 2017).

I show how territory, space, and resources in Greenland are (and have been) materialised, largely through trade, exploration, geological research, economic assessments, securitisation, assertions of sovereignty and political action, including self-government. In doing so, I explore the nature and place of territory and the subterranean in political and economic narratives and consider how the mapping, shaping and making of Greenland as territory, society, nation (and a hoped-for independent country) has been – and continues to be – bound up with processes and formations of enclosure, unearthing, and with the extraction of resources, whether they have been from the sea, land, or the subsurface. Enclosure in Greenland has various forms. The island has always been viewed as extractive terrain – or, more accurately, during a period of intensive whaling, the colonial trade era, and the more recent period of commercial fisheries and seismic surveys and offshore oil exploration, an extractive waterscape (cf. King 2020). And while Greenlandic political discussion on the shaping of Greenlandic society, nation, and (future) state, is entangled with discourses on the potential and possibility of wealth emerging from geological formations, Greenland's rapidly evolving political landscape, its economic development trajectory, and the politics concerning subterranean territory, including the seabed and the continental shelf, play a critical role in Danish-Greenlandic relations and have implications for the future of the Kingdom of Denmark as an Arctic state.

Although focused on Greenland, this book has a broader relevance and my hope is that it contributes to debates elsewhere in the Arctic and beyond. I locate Greenland within global discussion about extraction, the environmental and social impacts of resource development and sustainability, but also give attention to the place of the subterranean and geo-security in these debates. As I show, while ocean depths and subsurface geologies are essential to political narratives about greater autonomy and the economy, Greenland has long been a securitised space. This was especially so during the Cold War, but its strategic importance continues to be a matter of defence policy for the Kingdom of Denmark as well as Greenland's own emerging politics related to securitisation, and it influences the choices both Denmark and Greenland make about foreign policy. Greenland's air space, surface terrain, and underground are key to Arctic strategy making and are integral to matters of international relations, especially with the US, as well as bilateral agreements with other countries.

While touching on these geopolitical dimensions, my discussion does not lose sight of how Greenland's state formation ambitions have social, economic, and environmental implications. At one level, as Greenland moves through a post-colonial process of nation-building and state formation towards greater autonomy within – as well as political and economic distance from – the Kingdom of Denmark, it entails the possibility of different kinds of dependencies and influences. These will affect politics, economy, and society as Greenland becomes more deeply embedded in a global supply chain for raw materials, expands its business arrangements in North America and Asia, or looks to the US for funding for capacity-building and education. On another level, planning for the activities of extractive industry involves political procedures and volumetric practices that are used to map geological strata and the seabed and define and regulate resource spaces, all of which have social and cultural consequences. But what are designated as resource spaces are far from empty of long historical or contemporary human presence and social and cultural meaning. This has been a recurring point in some of my recent writing, in which I discuss how places that are inscribed as resource zones and in the process turned into mineral reserves are rich, lively worlds of past and present Inuit societies and intricate human-animal and broader human-environment relations. Nonetheless, they are often represented in environmental and social impact assessments (EIAs and SIAs), reports, and public hearings as remote wilderness areas.

I began this introduction with a discussion of climate change. While I consider some of the social and environmental effects on Greenland in several parts of this book, my critical departure is to go beyond a conventional description and chronicling of melting ice, environmental change, and Greenland's domestic and international politics. Again, as I have already mentioned, this is not a study of the impacts of a particular mine or mines either. I explore how Greenland's underground and its resources – and the associated infrastructure needed and the markets that are being targeted – matter for imagining the country's future, for the formulation of economic and development policy, for environmental governance, and for geo-security.

Combined with concern over climate change, the unearthing of resources from places that are being turned into mineral reserves highlights Greenland's geo-strategic position and its global entanglement, as various economic and political interests look to invest, establish business relations, and forge diplomatic links with Nuuk. This draws attention to thinking about historic and contemporary processes driven by a concern to ensure the geo-security of Greenland and the wider Arctic and North Atlantic. Here, I think of geo-security not just as something concerned with the nature of geographical relations between countries, such as the provision, maintenance, closure or openness of borders, for instance, or with the securitisation of state space or foreign affairs. I explore it in terms of what it offers for opening a much more expansive view of territoriality in which surface terrain, the subsoil, sea, and the atmosphere as well as human entanglements with the non-human are vital to Greenlandic autonomy, Indigenous sovereignty, geopolitical realignments, energy policy, and conservation. As Greenland positions itself globally, it also seeks to secure its own resources, take control of the decisions it makes about how they will be extracted and managed, and also how Greenlanders can geo-secure their future.

Greenland has a long history of being affected by colonial and imperial ventures, has been connected and tied tightly to global economic networks, and assumed geopolitical significance in the nineteenth and twentieth centuries. In Chapter 1, I relate some necessary history. In the eighteenth century, Denmark established a trade monopoly and Greenland became subject to Danish colonial rule – and to strict limits on who could travel there and exploit its resources – until 1953. During this time of territorial enclosure, merchants oversaw the development of trading stations and factories known as colonies that gathered seal and whale oil, skins, furs, and fish which were conveyed annually by ship to Copenhagen. Disciplined as Danish subjects and nurtured to be the producers of trade items, Inuit were converted to Christianity and were drawn into situations of social and economic dependency on these trading stations. Politically subservient to the Danish administrators, a stratified society emerged with a Greenlandic élite that discussed issues of identity and forged ideas about a Greenlandic nation from the nineteenth century onwards. During the early twentieth century, administrators began to imagine the development of a different kind of resource frontier. The chapter also discusses how Denmark extended its territorial claims over the entirety of Greenland, how Inuit were caught up in plans for asserting Danish sovereignty, and how ideas of geology and historic occupancy were central to this.

Chapter 2 examines the policy processes imposed by the Danish state on Greenland following World War II, especially during the 1950s and 1960s. I discuss how the ending of colonial rule marked the beginning of an era that was characterised by far-reaching economic transformation, infrastructural transfiguration, and profound social and cultural upheaval. This resulted from a process of resettlement and urbanisation, but this also spurred a critical response, as young Greenlanders formed political parties and began to call on the Danish state to grant Greenland self-government. I trace the beginning of this political movement, the achievement of Home Rule in 1979, and the nurturing of a stronger form of self-government with the granting of Self-Rule in 2009. I discuss how Home Rule was a process of nation-building that allowed for devolution, whereas Self-Rule has moved the nation-building process along a trajectory of self-determination towards state formation. Key to this is Greenland's ownership of subsurface resources and the implementation of government strategies for the development of extractive projects. With aspirations for greater autonomy and state formation, Greenland's political leaders have been mapping out trajectories for economic development, a stronger say in foreign policy, and possible independence. This is most recently marked by work towards the draft of a Greenlandic constitution.

In Chapter 3, I draw largely on my time in Nuuk to examine the nature of extractive industry in Greenland, which I suggest is largely a matter of speculation. Political geology matters in the making of a subterranean frontier and the regulation of extractive terrain. The chapter discusses how the anticipatory politics concerning a future Greenland (and various imaginaries of what that future could look like) play out in Nuuk as a centre of Greenlandic politics and business. I also discuss how the government of Greenland has recently made pronouncements that, while mining will continue to be one strong pillar for economic development, it will focus on sustainable mining as part of a green agenda for contributing to renewable energy. As attention turns to thinking about Greenland as a rare earths frontier, ensuring access to critical raw materials necessary for high-technology applications, including renewable energy development and the manufacture of consumer electronics, has become a matter of economic importance and strategic security as global demand increases. The final part of the chapter discusses how Nuuk is being remade; in emerging as an Arctic metropolis, it is undergoing a process of unbecoming as its colonial and post-colonial traces are being erased in order for the city to be refashioned as a dynamic, globally-connected capital.

Chapter 4 focuses on how discovering what lay under the ice, the land, and the sea bed also became central to much scientific research – a great deal of which has not just been carried out by Denmark, international scientists, and by mining and oil companies, but by the US during the Cold War. I discuss how Greenland became a securitised space and how understanding ice and terrain became essential to this process. During World War II, Greenland was vital for military transportation, aircraft movement, and weather forecasting; during the Cold War, it was central to northern hemispheric security concerns. Today, Greenland continues to figure prominently in contemporary discussions about security and surveillance in the Arctic (just as it has become key to global concern over climate change, resource development, and the economic and political calculations of future critical raw material production and trade).

The defence area of Thule Air Base, which was renamed Pituffik Space Base in April 2023, remains critical territory for the US, but it also remains crucial for Danish-American relations. As an Arctic state, it is through Greenland that the Kingdom of Denmark is able to position itself as a key player in world affairs. Danish relations with the US must be understood in part through historical and current US geopolitical interests in Greenland and how far Denmark is able to influence America's position on the Arctic. But Greenland's rapidly evolving political landscape, its economic development strategy, and its aspirations for independence now play a critical role in Danish-Greenlandic relations as well as for the forging of bilateral relations between Greenland and the US. However, I discuss how the securitisation process initiated during World War II, which intensified during the Cold War, had an impact on the Indigenous communities of Northwest Greenland and how this legacy is experienced today as one of dispossession.

Drawing from my long-term ethnographic work in Northwest Greenland, Chapters 5 and 6 illustrate how Greenland's resource spaces – and the extractive terrain and extractive waterscapes that have been marked out within them – have become sites entangled with wider global processes. Mining companies promote their projects by emphasising the uniqueness of Greenland's geology and the spectacular nature of the mineral deposits to be found there and that mines will create employment and significant economic opportunities. This is something that is welcomed by municipal authorities and the self-rule government alike. In Chapter 5, I discuss the regulatory and administrative nature of extractive industry in Greenland today. I also consider how the presence and influence of extractive industries in Greenland, especially in the surroundings that people depend on for their hunting and fishing livelihoods, result in contentious issues that have come to dominate national discussion. Amongst these, the nature of public participation in decision-making processes concerning the development of subsurface resources has been particularly controversial. Many people, for example, do not feel that they are sufficiently informed or consulted about the potential impacts of oil, gas, and mineral extraction projects, even though the Greenland government's more recently updated and revised guidelines for preparing SIAs and environmental impact assessments (EIAs) now have a clearer emphasis on a consultation process with affected communities and other stakeholders than their earlier versions outlined. The chapter also explores how the contemporary practice of geological exploration and research in Greenland is tied in with the mapping of territory, the geospatial assembling of resources, and the political discussion of the economic possibilities of extraction. I argue that remoteness matters in how resource spaces are scoped out, represented, and made, and illustrate this with a discussion about Greenland's far northern mine.

In Chapter 6, I continue to explore how many people feel they are not sufficiently informed or consulted about the potential impacts of extraction projects and that local views and concerns are not adequately included in SIAs and EIAs. Controversy surrounds many projects, and by drawing on my research in Northwest Greenland, I show how the technologies and economies of extraction are mobilised and enacted in resource spaces that are imagined as ontologically distinct from society and depicted as remote and empty of Indigenous human presence, even as they are assembled and constituted as sites of potential for transforming Greenlandic society. Northwest Greenland has long been viewed as a resource space by Indigenous hunters and more recent arrivals, including explorers, whalers, traders, geologists, and the mining and oil industries. People living in the coastal communities of the Upernavik and Avanersuaq districts call for greater inclusion of Indigenous and local perspectives, and I discuss this with reference to mineral exploration and seismic surveys. I suggest that an attentiveness to community perspectives on human-environment relations as well as Indigenous knowledge and insights on the non-human would be a major step forward for how resource companies conduct EIAs and SIAs.

Chapter 7 discusses how Arctic sea ice, glaciers, coastlines, and seas have become sites and objects for new forms of environmental governance that are shaped by ideas of unique and fragile ecosystems under threat in the Anthropocene. This nurtures a spatial politics of disappearance sustained by scenarios of ice-free Arctic futures and informed by ideas of the spectacular and worries over extinction. Conservation organisations frame the Arctic as a zone of climate change crisis and have launched campaigns – underpinned by narratives of ruination – to protect 'last areas' of ice or ecologically sensitive waters such as High Arctic polynyas. The chapter shows, however, that Inuit responses to the possibility of resource extraction as well as environmental protection formulated without their participation challenge and seek to resist environmental governance and conservationist narratives that do not necessarily take note of Indigenous perspectives on place and human and non-human relations. Along with the world's surface, the subterranean, the ocean depths, and the atmospheric are enmeshed with the lives of humans, animals, and other non-human entities. Indigenous people's organisations, such as the Inuit Circumpolar Council, argue that the need to ensure the inclusion of local communities in conservation initiatives for marine environments, ice, and wildlife acquires added urgency, given the speed of climate change, the international interest in Arctic resources, and increased shipping.

A note on names, nomenclature, and terminology

Greenland is geographically part of the North American continent, but along with Denmark and the Faroe Islands, it is a constituent part of the Kingdom of Denmark (*Kongeriget Danmark* in Danish). The Danish Realm (*Rigsfælleskabet*) is the name used to refer to the relations between the three constituent parts of the kingdom, which are also often referred to as countries, although it is more accurate politically and constitutionally to refer only to Denmark as a country and to Greenland and the Faroe Islands as self-governing territories. However, as I discuss in Chapter 2, Greenlanders are also recognised as a nation. In this book, I refer to Greenland as both a territory and a country. Greenland has a population of some 57,000, of whom 88% are Inuit, or who identify themselves as ethnic Greenlanders as opposed to those who identify themselves as Danish or who are from other countries. Greenland's Inuit population comprises three distinct cultural and linguistic groups: the majority Kalaallit, living along the west coast from Nanortalik district in the south to the Upernavik area in the north; Inughuit (popularly known as Polar Inuit and famous as the world's most northerly Indigenous inhabitants) in the north around Avanersuaq/Thule; and Iivit on the east coast in the Tasiilaq and Ittoqqortoormiit areas.

Since Home Rule was introduced in Greenland in 1979, the country has been known officially as *Kalaallit Nunaat* ('the Greenlanders' Land'), although both 'Inuit' and 'Greenlanders' are used as more generic and interchangeable terms to refer to the Indigenous population. 'Inuit' (sing. 'Inuk') means 'people' (or 'human beings') in Kalaallisut, as it does in Inuktitut, the Indigenous language of Nunavut. Despite the regional distinctions, Kalaallit (singular Kalaaleq) is also used to refer to – and for people to identify themselves as – Greenlanders, wherever they happen to live in Greenland, or in Denmark, or elsewhere. Danes are referred to in Greenlandic as Qallunaat (singular Qallunaaq). It should be noted however that 'Greenlander' is also applied to those who can identify and be identified as someone who is ordinarily resident in Greenland, including Danes. This can blur ethnic boundaries, and the politics of identity in Greenland often centres – indeed increasingly so – on ideas and notions of Inuit identity to distinguish those who identify as Indigenous people vis-à-vis other non-Indigenous residents. Greenlandic is an agglutinative language. The suffix *-miut* (singular *-mioq*) means 'person of' or 'inhabitant of' and is used to indicate identity, belonging, or residence of a place (depending on the rules of word formation, this is sometimes *-mmiut* or *-mmioq*). For example, someone from Nuuk is a Nuummioq (plural Nuummiut), while someone from Upernavik is an Upernavimmioq (plural Upernavimmiut). Kalaallisut is the official language of Greenland, although Danish is widely used as a language of administration. In different parts of the text, I use both 'Kalaallisut' and 'Greenlandic' when I refer to the Indigenous language.

Finally, and as ever, there are a number of colleagues, friends, institutions, and funding agencies to whom I owe much thanks. I am grateful to the support given to me by the Department of Anthropology at the University of Alberta and for funding from the Henry Marshall Tory Chair research programme. In Nuuk, Ilisimatusarfik/University of Greenland and the Greenland Climate Research Centre (GCRC) at the Greenland Institute of Natural Resources (GINR) have, over a number of years, generously provided space and institutional support. This book also draws on research undertaken through the Climate and Society programme, with funding from GCRC and GINR and the Danish Agency of Science, Technology and Innovation (Project 6400). Some chapters also draw on research in Northwest Greenland that was funded by the EU Framework 7 project ICE-ARC (Ice, Climate and Economics: Arctic Research on Change, grant number 603887). In Greenland, there are, as always, too many people to thank. My late colleague and friend Lene Kielsen Holm at GCRC will be missed and I will always reflect on how privileged I was to know her and to work with her as we built the Climate and Society programme together in Nuuk and carried out research in Nuup Kangerlua and in Avanersuaq. To Vittus Nielsen, Kaaleeraq Tobiassen, Marius Tobiassen, and Angunnguaq Josefsen, thank you for the work we did in partnership together and for all the travels we enjoyed. Marius and Angunnguaq will be similarly missed. Also in Nuuk, thanks to Birger Poppel, MariKathrine Poppel and Upaluk Poppel, and to Michael Schluchtmann. In Northwest Greenland, I will always owe a debt of gratitude to David and Birthe Kristiansen, Godman and Judithe Jensen, Mamarut Kristiansen, Tukumeq Peary and Qultana Qvist, and to so many others. At Routledge, my thanks to Hannah Ferguson and Katie Stokes, and Jennifer Hicks, for their support and patience as this book has taken shape. And, as always, I am so grateful to Anita Dey Nuttall and Rohan Nuttall for the encouragement they give me. A mere thank you, I feel, will never do justice to what that means.

Figure 0.1 Greenland. Attribution: Ian Macky, 2014 PAT Atlas. Public domain, via Wikimedia Commons.

Source: https://ian.macky.net/pat/map/gl/gl_blu.gif

References

AMAP. 2012. *Arctic Climate Issues 2011: Changes in Arctic snow, water, ice and permafrost.* Oslo: Arctic Monitoring and Assessment Programme.

AMAP. 2017. *Snow, Water, Ice and Permafrost in the Arctic: Summary for policy-makers.* Oslo: Arctic Monitoring and Assessment Programme.

Andresen, Camilla S., Kristian K. Kjeldsen, Benjamin Harden, Niels Nørgaard-Pedersen and Kurt H. Kjær. 2014. "Outlet glacier dynamics and bathymetry at Upernavik Isstrøm and Upernavik Isfjord, North-West Greenland." *Geological Survey of Denmark and Greenland Bulletin* 31: 79–82.

Arboleda, Martín. 2020. *Planetary Mine: Territories of extraction under late capitalism.* London and New York: Verso.

Arsel, Murat, Barbara Hogenboom and Lorenzo Pellegrini. 2016. "The extractive imperative in Latin America." *The Extractive Industries and Society* 3 (4): 880–887.

Ballinger, Thomas J., G.W.K. Moore, Yarisbel Garcia-Quintana, Paul G. Myers, Amreen A, Amrit, Dániel Topál and Walter N. Meier. 2022 "Abrupt northern Baffin Bay autumn warming and sea-ice loss since the turn of the twenty-first century." *Geophysical Research Letters* 49 (21): e2022GL101472, https://doi.org/10.1029/2022GL101472.

Bjørst, Lill Rastad. 2020. "Stories, emotions, partnerships and the quest for stable relationships in the Greenlandic mining sector." *Polar Record* 56, E23. doi: 10.1017/S0032247420000261.

Box, Jason E., Alun Hubbard, David B. Bahr, William T. Colgan, Xavier Fettweis, Kenneth D. Mankoff, Adroen Wehrlé, Brice Noël, Michiel R. van den Broeke, Bert Wouters, Anders A. Bjørk and Robert S. Fausto. 2022. "Greenland ice sheet climate disequilibrium and committed sea-level rise." *Nature Climate Change* 12: 808: 813.

Carse, Ashley and David Kneas. 2019. "Unbuilt and unfinished: the temporalities of infrastructure." *Environment and Society: Advances in Research* 10: 9–28.

Dodds, Klaus and Mark Nuttall. 2016. *The Scramble for the Poles: The geopolitics of the Arctic and Antarctic.* Cambridge: Polity.

Dodds, Klaus and Mark Nuttall. 2018. "Materialising Greenland within a critical Arctic geopolitics." In Kristian Søby Kristensen and Jon Rahbek-Clemmensen (eds.) *Greenland and the International Relations of a Changing Arctic: Postcolonial paradiplomacy between High and Low Politics.* London and New York: Routledge, pp. 139–154.

Elden, Stuart. 2013. "Secure the volume: vertical geopolitics and the depth of power." *Political Geography* 34: 35–51.

Espig, Martin and Kim de Rijke. 2018. "Energy, anthropology and ethnography: On the challenges of studying unconventional gas developments in Australia." *Energy Research and Social Science* 45: 214–223.

Hastrup, Kirsten. 2009. "Arctic hunters: climate variability and social flexibility." In Kirsten Hastrup (ed.) *The Question of Resilience: Social responses to climate change.* Copenhagen: Royal Danish Academy of Science and Letters, pp. 245–270.

Hein, Carola. 2018. "'Old refineries rarely die': Port city refineries as key nodes in the global petroleumscape." *Canadian Journal of History* 53 (3): 450–479.

King, Owen. 2020. "Tight oil and water: Climate change and the extractive waterscapes of Western Siberia." In Regina M. Buono, Elena López, Jennifer McKay and Chado Staddon (eds.) *Regulating Water Security in Unconventional Oil and Gas.* Cham: Springer, pp. 175–195.

Labban, Mazen. 2014. "Deterritorializing extraction: Bioaccumulation and the planetary mine." *Annals of the Association of American Geographers* 104 (3): 560–576.

Meredith, Michael and Martin Summerkorn 2019. "Chapter 3: Polar Regions." In H.-O. Pörtner et al. (eds.) *IPCC Special Report on the Ocean and Cryosphere in a Changing Climate*. Cambridge: Cambridge University Press, pp. 203–320.

Muilwijk, Moven, Fiamma Straneo, Donald A. Slater, Lars A. Smedsrud, James Holte, Michael Wood, Camilla S. Andresen and Ben Harden. 2022. "Export of ice Sheet Meltwater from Upernavik Fjord, West Greenland." *Journal of Physical Oceanography* 52 (3): 363–382.

Nuttall, Mark. 1992. *Arctic Homeland: Kinship, community and development in Northwest Greenland*. Toronto: University of Toronto Press.

Nuttall, Mark. 2008. "Climate change and the warming politics of autonomy in Greenland." *Indigenous Affairs* 1–2/08: 44–51.

Nuttall, Mark. 2017. *Climate, Society and Subsurface Politics in Greenland: Under the Great Ice*. London and New York: Routledge.

Nuttall, Mark. 2022. 'Places of memory, anticipation, and agitation in Northwest Greenland' in Kenneth L. Pratt and Scott A. Heyes (eds.) *Memory and Landscape: Indigenous responses to a changing North*. Athabasca: Athabasca University Press, pp. 157–177.

Parr, Joy. 2010. *Sensing Changes: Technologies, environments, and the everyday 1953–2003*. Vancouver: University of British Columbia Press.

Priebe, Janina. 2017. *Greenland's Future: Narratives of natural resource development in the 1900s until the 1960s*. Umeå: Umeå Universitet.

Rasmussen, Mattias Borg and Christian Lund. 2018. "Reconfiguring frontier spaces: the territorialization of resource control." *World Development* 101: 388-399.

Schritt, Jannik and Andrea Behrends. 2018. "'Western' and 'Chinese' oil zones: Petro-infrastructures and the emergence of new trans-territorial spaces of order in Niger and Chad." In Ulf Engel, Marc Boeckler and Detlef Müller-Mahn (eds.) *Spatial Practices: Territory, Border and Infrastructure in Africa*. Boston: Brill, pp. 211–230.

Sejersen, Frank. 2020. "Brokers of hope: Extractive industries and the dynamics of future-making in post-colonial Greenland." *Polar Record* 56. doi.org/10.1017/S0032247419000457.

1 Enclosure and extraction

If there is one thing that has been consistent throughout Greenland's recent history, it is that enormous effort has gone into political strategy, policy, resource extraction, design, construction, and engineering to make and reproduce Greenlandic society and economy as well as get to know its environment, assess and exploit its resources, classify and control (and assimilate) its Indigenous inhabitants, and determine its geographical extent. Greenland has often been a place on which Danish dreams have been projected and in which Danish government policies – often experimental, misguided, and misjudged – of territorial enclosure, enclaving, resource extraction, state planning, a 'civilising' process, and social engineering have been enacted. Greenland exemplifies what Jill Franks (2006: 1) describes in her writing about the allure of islands – distant from a mainland, they are often thought of as isolated, but at the same time they are considered to be controllable, yet imagined as paradisiacal. Islands figure as testing grounds, as social laboratories, or they are sites of fantasies, utopia, and personal transformation.

During the era of Denmark's colonial ventures, which were underpinned by the commodification of marine mammals, fish, and other animals and the trade of the products that resulted from their exploitation, ideas about a particular form of an 'ideal' and 'genuine' Greenlandic culture and way of life based on hunting (and the hunting family unit) were reinforced and reproduced. This was done so that resources could be extracted from the non-human beings that inhabit sea and land. During World War II and throughout the Cold War (as I discuss in Chapter 4), American military planners and strategists went about getting to know the properties of ice, surface terrain, and the seabed, and sought to design airstrips, military bases, weather stations, and subsurface structures as Greenland was made into a geostrategic space for security and surveillance. Following the end of World War II, Danish policy was directed towards the design of a new, modern Greenlandic society. In the 1970s, Greenlandic politicians reacted to the social and economic turbulence and transformations that resulted and began to push for self-government, which resulted in Home Rule being enacted in 1979. They have since sought to design a Greenlandic nation and lay the foundations for an independent state – a greater degree of autonomy was achieved with Self-Rule in 2009, which recognises Greenlanders as a nation with the inherent right to independence if they choose it. Today, Greenland is viewed as a new resource frontier for extractive industry. It

DOI: 10.4324/9781003175421-2

is also seen by planners and architects as a place that is a blank slate for innovation in technology and engineering, architecture, and new forms of Arctic living and urban design as colonial and post-colonial traces are erased.

This book discusses all this, and to begin with, in this chapter, I provide some necessary coverage of the colonial history of Greenland as a form of enclosure and resource extraction. By this I do not mean enclosure as a form of appropriation and the carving up and privatisation of land or territory within Greenland to allow for settler colonialism, but as a way of creating, making, administering, and securing Greenland as a clearly bounded territory – a resource space and later, a resource frontier – overseen and controlled by the Danish state for the purpose of extracting living marine and terrestrial resources to support a flourishing trade economy. This also involved the making and enclosure of lives, occupations, and subjects, and the exclusion of others who had designs on trade in Greenland or who sought to extract what were categorised as marine resources such as whales or trap Arctic foxes for their furs. Greenland gradually emerged as a closed territory – an exclusive Danish enclave even – through a process of geo-security that was administered by Danish authorities over a jurisdiction which they sought to ensure other countries had no claim or rights of access to.

Imaginings

Greenland has long been imagined by those who do not live there as a remote Arctic territory, a far-flung, icy island that is peripheral to global affairs. From the late sixteenth century, European explorers who embarked on discovery expeditions to the Arctic in search of a northwest passage to Asia through the northern waters of North America returned home with fantastic tales of Greenland and its Indigenous people. It was romanticised as a wild and unforgiving place, and books, narratives of northern voyages, and paintings of Arctic landscapes captured the public imagination. When English navigator John Davis viewed the southern coast on his first voyage to the Arctic in 1585, he called it "the land of desolation." Davis described mountains covered in snow, land that was deformed and rocky, and how ice was everywhere he looked, making an "irksome" noise. Ship's surgeon Bernard O'Reilly's account of a voyage to West Greenland on the whaler *Thomas* in the summer of 1817 characterised the coast as nothing but "barren rock" with a "dreary appearance." Mountains presented for him "the most dismal and chilling sight in nature" (O'Reilly 1818: 29–30). Almost 300 years after Davis sailed north, American explorer Isaac Israel Hayes borrowed his description of the south coast to use as the title of a book published in 1871. In *The Land of Desolation: A personal narrative of adventure in Greenland*, Hayes wrote how, on "a gloomy night" in July 1869, the ship *Panther*:

> fitted out for a summer voyage by a party in pursuit of pleasure, came in like manner, through a thick and heavy mist, to a place where there was a mighty roaring as of waves dashing on a rocky shore.
>
> (Hayes 1871: 2)

The ship, Hayes and his fellow passengers soon discovered, was "embayed in fields and hills of ice" and those on board had their first glimpse and experience of Davis's land of desolation. It was:

> A mysterious land to them, and one around which clung many marvelous associations. Its legends had been the wonder of their boyhood; its grandeur was now their admiration. They had heard of it as a land of fable; tradition had peopled it with dwarfs and giants; history recorded that a race of men once occupied it whose fleets of ships traversed the waters in which their own vessel was now so grievously beset, bearing merchandise to hamlets of peace and plenty. Their eyes naturally sought a spot whereon to locate the home of this ancient people; but nothing could they discover save sterile rocks and de-sert wastes of ice. They saw dark cliffs which rose threateningly above them abruptly from the sea, and beyond these their eye wandered away into the interior, which the snows of centuries had converted into a vast plain of desolate whiteness.
>
> (ibid.: 2–3)

Hayes went on to recount the disconcerting experience of listening to ice that creaked and groaned, gazing upon icebergs of "enormous magnitude," and sailing through "angry and troubled waters." This description of the grandeur and power of Arctic phenomena is characteristic of nineteenth-century accounts of polar travels, when explorers continued to search for the Northwest Passage, looked for routes to the North Pole via Baffin Bay and Nares Strait – the waterway between Ellesmere Island and Greenland – and when whalers sailed north in their wooden vessels along Greenland's west coast fearful of the pack ice and its dangers. Greenland continued to inspire feelings of awe, mystery, and dread – what English geographer Clements Markham (1853: 1) called a "strange admixture of the sub-lime and terrible in the stern unbending landscape." While nineteenth-century Arctic explorers were regarded as witnesses to the sublime and the terrifying and were celebrated at home for their heroic feats in an icy wilderness, there was also a ready audience eager to hear and read about disaster stories that told of ships crushed by ice and entire expeditions getting lost in a terrible white emptiness (Behrisch 2003; Craciun 2011). Similarly, artistic representations of the Arctic such as Edward Landseer's *Man Proposes, God Disposes* (1864) and John Macallan Swan's *The Abandoned Boat* (c. 1879–1910) imagined scenes of terror, isolation, and tragedy in an unforgiving environment, while Briton Rivière's *Beyond Man's Footsteps* (1894) presents a sublime scene of pack ice, one that is empty of people but in which the polar bear is majestic.

Indigenous people are often noticeably absent from accounts of the exploration of what were described by Europeans and North Americans as unknown Arctic spaces. When their presence was acknowledged, they were more often than not considered to be scientific curiosities subsisting from meagre resources in a bleak environment. There was often no interest in learning from them about how to survive in the Arctic or with chronicling the knowledge they had about their surroundings that would have assisted explorers in finding their way across the ice, through the northern seas, and over the terrain they often got lost in. Although Lutheran priest Hans Egede and Moravian missionary David Crantz had published descriptions of Greenland and its people in the eighteenth century – with a focus on lifeways, material culture, spiritual beliefs, and hunting practices – and even though by then the trade economy had led to social stratification and occupational differentiation, Inuit were nonetheless often portrayed in the annals of discovery and the journals of whaling voyages as a primitive people – indeed, depicted as other-than-human – without their own modernity and a history of deep global connection. Hayes describes an encounter in South Greenland in this way:

> Presently we saw something dark moving upon the water, which appeared to have the body of a beast, and the head and shoulders of a man. It might be a marine centaur! who could tell? In fact, we rather expected to see some such monsters long before; and if the sea had been alive with them, we would not have been, I think, much surprised.
>
> (ibid.: 6)

This turned out to be the pilot from Qaqortoq (then known as the Danish trading colony of Julianehåb) approaching in a kayak to guide them into the harbour. He climbed on board and once on deck, Hayes described his appearance:

> His body was covered in hair, and he was all wet, as if he had just risen from the bottom of the sea. Besides, he smelt fishy. Yet this was clearly the best we could do if we ever meant to get into port....
>
> (ibid.: 7)

Hayes thought Qaqortoq to be flourishing because of the Danish trade activities there, but called it a hopeful town in a hopeless place – the residents 'emerged' from huts that were "scarcely distinguishable from the rocks themselves, and the people appeared to be coming out of the earth, and dropping into it again like prairie-dogs" (ibid.: 12–13), while further north, beyond Upernavik at a latitude of 72°, he said "there are no Christian people, or people of any kind living on the earth, except a few skin-clad savages" (ibid.: 12).

Such impressions about Greenland and its people were difficult to dispel even during much of the twentieth century. When A.C. Seward, who was professor of botany at the University of Cambridge, published an account of a summer visit to the country made in 1921 in search of fossilised plants (he also described Greenland's geology and geographical characteristics), he related his feeling that it was:

> a land which in some respects merits the name given to it more than three hundred years ago—the Land of Desolation; it is a land remarkable for the splendid dignity of its scenery and possessed of a subtle power of inspiring affection tempered by a sense of awe.
>
> (Seward 1922: 98)

In 1951, sociologist and political scientist Joseph Roucek thought of Greenland as "an enormous hunk of ice" where only a few people "scratch out an existence" (Roucek 1951: 239), while in 1953, Geoffrey Williamson wrote that "A modern traveller, arriving in Greenland by air, is quite likely to imagine that he is landing on another planet" (Williamson 1953: 3). For Roucek, although Greenland was strategically and geopolitically important to the US, it was still an "inhospitable territory."

Despite the political significance and geostrategic location that Roucek emphasised, Wayne Fisher, who was the last diplomat to serve as US Consul for Greenland at the original American mission in Nuuk (then officially known as Godthåb) when it closed in 1953, thought it "one of the strangest posts in the entire American Foreign Service" (Fisher 1954: 30). In an account of the final days and closure of the consulate, which was opened in 1940 following Germany's invasion of Denmark, Fisher wrote about the self-reliance needed to operate it as well as maintain the building in which it was housed and the isolation he and his wife experienced during their posting:

> No couriers ever came to Greenland. In winter my outgoing pouch mail would sometimes wait as much as two months for some means of dispatching it, and even then often had to be forwarded to Washington via Copenhagen, on a Danish ship calling at Godthaab. The difficulty of travel not only to and from Greenland but also within the country, which has no railroads or highways whatsoever and no dogs in the southern half, is hard to imagine for one who has not been there.
>
> (Fisher ibid.: 52)

"To top all this," he went on to write:

> my wife and I once went seven full months without even seeing any other Americans. Incredible as this may sound in these days when the American Government with all its attendant and elaborate supply system is operating in every nook and cranny of the earth, it nevertheless happened.
>
> (ibid.)

And he had this to say about the Indigenous Greenlanders:

> There is nothing fawning or servile in their friendship, since there are no more independent people anywhere in the world. Some outsiders might find their nature rather roughhewn. True, the Greenlanders' sense of humor can be a bit rough. If, for example, someone struggling along with two bucketfuls of water slips and falls on the ice, an onlooker invariably breaks into a loud guffaw, whether the unfortunate one be male or female. I believe this feeling, however, stems from a "survival of the fittest" attitude spawned by the traditional difficulty of survival in the barren land of ice and rocks which contains practically no vegetation of any kind. Their attitude toward animals

and game birds might also seem rather harsh to outsiders, but it occurs to me that if our environment were such that survival itself could depend on our success in hunting, we might not be as inclined toward a "be kind to animals" philosophy either.

(Fisher ibid.)

These kinds of descriptive accounts ignored Greenland as a place of lively encounters between human and non-human entities, where Inuit have lived for some 4,500 years dependent on marine and terrestrial animals, and where lo-cal, national, and global connections, forces, influences, and assemblages have gathered, coalesced, and collided (Dodds and Nuttall 2018). Representations of remoteness, ice, wilderness, and the sublime influenced public perceptions about Greenland – and indeed, the wider Arctic – and they still play a critical role in shaping contemporary views. Hanrahan (2017: 103) writes that depictions of Greenland and Greenlandic Inuit in English language explorer accounts, in the period from the early 1870s through to the 1920s, reinforced "the Arctic imagi-nary that helped establish Greenland's ongoing subordinate position in the global economy" and hindered efforts to realise Greenlandic self-determination. In the later part of the nineteenth century and early decades of the twentieth, the re-ports and narratives of explorers were critical for informing governments about the importance of the Arctic for defence and for preliminary assessments of its value as a frontier for the extraction of minerals (Fogelson 1985), but they also in-fluenced political ideas and business interests about the Arctic as empty, making it easy to ignore the knowledge and agency of Indigenous people and to set about planning to unearth its subsurface resources (Hanrahan ibid.).

Today, most travellers arrive in Greenland by air. Some tourists go there by cruise ship, but for many, whether they are returning home or visiting for the first time, the first glimpse of the world's largest island is the east coast and the inland ice from the windows of Air Greenland's one airliner, an Airbus A330-200 named *Norsaq* (which is the name for the throwing board by which a harpoon is hurled by a hunter, usually from a kayak). This flies on weekdays from Copenhagen's Kastrup Airport over the northern North Atlantic to Kangerlussuaq on the west coast, which for the time being remains Greenland's main international airport. By 2024, both Nuuk, Greenland's capital, and Ilulissat – the country's third largest town and hub of the tourism industry in Disko Bay – will have new international airports open that will be able to accommodate transatlantic aircraft making di-rect flights from Copenhagen and, so politicians, business leaders, and Air Green-land's directors hope, from other major cities in Europe, North America, and Asia (*Norsaq* is also being replaced by an Airbus A330–800 named *Tuukkaq*, which means 'harpoon head'). This infrastructure is critical for ensuring Greenland, especially its capital city, has greater global connectivity. Air Greenland and Air Iceland also fly Dash-8 aircraft to Nuuk (Air Iceland flies seasonally to Ilulissat as well) from Iceland's international airport at Keflavik and its domestic hub in Rey-kjavik, respectively. However, the majority of air passengers currently make the journey – a four and a half hour, or so, flight – from Copenhagen. As the Airbus approaches Greenland's east coast, snow-capped mountains, ice-filled fjords, and coastal waters come into view. The aircraft then takes around another hour to reach Kangerlussuaq. Cloud cover means the inland ice disappears from view for a while, but as the Airbus makes its landing approach, it dips low, allowing a glimpse of an expanse of Greenland's white, icy interior. I have sat next to many people visiting Greenland for the first time for whom this reinforces their impression of arriving in a vast, remote, almost dreamlike, polar region.

As I show in this book, these impressions and representations of remoteness and unspoiled nature also matter to how Greenland is thought of by mining and oil companies as well as by conservationists and environmental organisations. I argue that Greenland continues to be thought of as remote and is often represented, spoken, and written about in terms of the spectacular (cf. Igoe 2017). This is evident in how it is described as a wilderness with breathtaking scenery, but with a fragile biodiversity; in the ways strategies are promoted for ecosystem protection and wildlife management and in the ways it is also abstracted as a new resource frontier with enormous and unique mineral reserves, including the critical minerals the world considers vital for a green energy transition.

The first peoples

Some 4,500 years ago or so, the first Paleo-Inuit groups, named by archaeology as the Independence I culture, arrived in Greenland by taking a northerly route by way of what is now known as Canada's Arctic region. With a cultural connection to the Arctic Small Tool Tradition, which originated at Cape Denbigh in Alaska, they moved across land, water, and ice from the Canadian High Arctic, in the areas that stretch from the North Water polynya (Pikialasorsuaq in Greenlandic; I write more about this vitally important and productive ecosystem in Chapter 7) and north to Nares Strait. They lived in Greenland between c. 2400 BCE and 1900 BCE, mainly in the northernmost part of Greenland, centred around Peary Land and the far north eastern coastal areas, with scattered locations extending to the region around Clavering Island. Venturing inland to procure their resources, they depended primarily on hunting large land mammals such as musk ox and reindeer, but also pursued seals and other marine mammals (Grønnow and Jensen 2003).

The Saqqaq culture – named after the settlement near Ilulissat in Disko Bay, which is the best known site of occupancy, is dated between c. 2300 BCE and 1000 BCE and is West and East Greenland's counterpart to the Independence I culture. Significant Saqqaq culture locations have been identified along the west coast from Upernavik in the north to Qaqortoq in the south and around the Tasiilaq region on the east coast, with a few sites north to Clavering Island (Møbjerg 1986). The Saqqaq people were hunters of seals and reindeer and also had cultural links with the Arctic Small Tool Tradition. The Independence II culture is the name given to a new wave of migrants who inhabited northern Greenland from around 700 BCE to 80 BCE. There are fewer known sites compared to Independence I, but those that have been located by archaeologists suggest a preference for settlement in coastal inlets rather than the outer coast (Grønnow and Jensen ibid.). The sub-Arctic Dorset I culture (dated from around 600 BCE to around 1 CE) succeeded the Saqqaq culture, while evidence of the Dorset II culture (with sites dated between c. 700 BCE and 1300 CE) is found primarily in northern Greenland (Møbjerg ibid.).

The Thule culture Inuit, who are the direct ancestors of Greenland's present-day Inuit, emerged as a distinctive people along the coasts of Alaska around 1000 CE and moved rapidly across Arctic Canada to Greenland in several waves (Friesen 2022). The Thule culture developed from interactions with other cultures in northern Alaska such as Birnirk and Punuk. They had a primarily marine-based subsistence culture, focused on hunting large marine mammals such as bowhead whales, and were experts at travelling and hunting by skin boats (the *qajaq* and *umiaq*, rendered in English as kayak and umiak) during summer and by dog sledge across sea ice in winter and spring. Their patterns of social organisation and their strategies and elaborate technologies for hunting were highly flexible and could be modified to the changing ecology of the Arctic and the movement and availability of animals as well as the new environments they moved into. They took advantage of a full range of resources – bowheads, seals, walrus, beluga, musk oxen, caribou, and various species of fish and migratory birds (Friesen ibid.: 24). However, it was the bowhead and walrus that allowed them to concentrate their hunting activities in coastal areas where these marine mammals were abundant. Some of the most important winter and summer sites can be found today in several places on the south coast of Devon Island in Nunavut, for example. During the lengthy northern winter, families tended to group together in semi-permanent settlements. Winter houses were semi-subterranean and made of stone and turf, with a frame of driftwood, whalebone, or reindeer antler. Thule culture Inuit also lived in Northeast Greenland from around 1400 to 1850 (Kroon and Jakobsen 2010).

While it is suggested in the literature that a combination of a warming climate, increasing populations, social conflict, and competition for resources in the Bering Sea region and northern Alaska (not only for marine and land animals, but metals too), as well as in the High Arctic during the Middle Ages, allowed for this expansion to the east from Alaska and western Canada and on to Greenland (see Mason 2020, for a review of the literature), work by archaeologist Robert McGhee (2000, 2009) has argued that one of the main reasons for the initial Thule movement across the Arctic to Greenland was also a search for meteoric iron. This was known to be located around the area of the present-day community of Savissivik in the Melville Bay area and was considered a critical and valued resource for use in making tools and hunting implements. Savissivik means "the place where one gets iron or material for knives" on account of the meteoric fragments found nearby. As Friesen (ibid.) discusses, Thule Inuit and their ancestors had access to iron and other metals through centuries-old trade networks that stretched across the Bering Strait to Asia. Iron, for instance, was vital for making tools and allowed for technological advances in the weapons required for hunting, especially large marine mammals such as the bowhead. Having access to and accumulating stores of iron and other metals such as copper likely contributed to social status, and so the seeking out of potential sources as far as the eastern Arctic and Greenland could have been one impetus for the Thule movement. As Thule Inuit travelled and participated in trade networks, it is likely they heard stories of metal-rich places to the east. They may also have come to know about iron and bronze trade goods that circulated in Greenland after the arrival of the Norse.

Greenland was first settled by the Norse around 985 CE, when Eirík the Red and a group of followers sailed from Iceland. They took cattle, sheep, and goats with them from their home country and established homesteads in coastal areas with fertile land. Norse farming and fishing settlements expanded in two areas – *Eystribyggð* (Eastern Settlement) in the south and *Vestribyggð* (Western Settlement) in the southwest, in the fjords around the present-day Nuuk region (Arneborg 2018; Madsen 2019; Seaver 1996). The Norse Greenlanders also hunted marine mammals – mainly harp and hooded seals – and went on long voyages to exploit resource spaces in Greenland's north-western areas, even perhaps crossing to the eastern coast of what is now Ellesmere Island in Nunavut, to hunt walrus (Arneborg 2021; McGovern 2017) for the ivory from their tusks, an item which became an essential part of trade between Greenland and Scandinavia and northern Europe. The ivory was also used as a form of tithe payments to the Catholic Church.

The Icelandic sagas record that the Norse population in Greenland knew of and came into contact with people they called *Skraelings*, possibly on these long hunting voyages to the far north as well as around the *Eystribyggð* and *Vestribyggð*. This is the earliest written record of contact between Europeans and Inuit in Greenland, and recent archaeological research suggests this contact was made by 1170–1200 (Arneborg 2003: 173; Dugmore et al. 2007: 17). Norse settlement was organised initially as a free state. Following the conversion of the Norse to Christianity, the Catholic Church appointed the first bishop of Greenland in 1124. In 1261, the Norse Greenlanders became part of the Kingdom of Norway and paid their taxes to the crown. When Denmark, Norway, and Sweden were united in 1397, Norway's possessions in the North Atlantic, including Greenland, also came under Danish rule.

The Norse settlements were active for nearly five centuries, but contact between them and the wider Norse North Atlantic world became tenuous and was gradually severed. The last recorded contact by ship was in 1410 (Seaver ibid.) and the settlements that remained were eventually abandoned in the middle of the fifteenth century (Vésteinsson et al. 2016). Archaeologists and historians have long debated the reasons why and have put forward a number of theories to account for the decline of the Norse Greenland settlements, from poor heath, malnourishment and starvation, inbreeding, conflict and hostile encounters with Thule Inuit groups over scarce resources, assimilation with the Thule culture, and failure to adapt to a changing climate (e.g. Seaver ibid.). Contemporary archaeological evidence dismisses the view that the Norse colonies were either wiped out or assimilated by the Thule Inuit, though, and suggests instead that the Norse and Thule Inuit engaged in intermittent trade rather than their relations being characterised by conflict (McGhee 1984; Schledermann 2000; Sutherland 2000).

When the Norse first arrived in Greenland, parts of the island had been experiencing relatively mild climatic patterns during a time that has been termed the Medieval Warm Period. Data from ice cores, taken from the Greenland ice sheet, shows that this warm period lasted from ca. 800 CE until ca. 1300 CE, a time frame which covers approximately 65% of the Norse settlement period in Greenland and was the period during which Thule culture Inuit moved into and around the coastal areas. However, after ca. 1300 CE, the climate began to fluctuate more often and gradually cooled into the Little Ice Age. This would have affected the migration routes of seal species also hunted by the Norse. The archaeological record indicates that the *Vestribyggð* declined around 1400 CE, and the *Eystribyggð* held on until around 1450 CE before also being abandoned (Arneborg 2012). This was also a time when Thule culture transitioned from a more coastal-based bowhead whaling complex to a way of life that was more focused on hunting seals on the sea ice. The Little Ice Age had consequences for the seasonal availability, the location and abundance of key species such as bowhead whales, and for human mobility, so the more fixed, concentrated, and populous Thule settlements were abandoned in favour of a form of social organisation that was smaller in scale, nomadic, flexible, and dependent on sea ice.

Arneborg (2021) points out that walrus hunting and a profitable trade in ivory with Scandinavia and other parts of Europe were central to the social and economic life of the Norse settlements in Greenland. Walrus ivory was greatly coveted as a commodity in western Europe and Russia, and from the early part of the tenth century, Novgorod merchants assumed greater control over the hunting and trading of Russian Arctic products and focused on supplying lucrative markets in the Near East and Asia. This affected the availability of ivory supplies to western Europe, which came to rely on walrus hunting grounds in the North Atlantic. Arneborg and other archaeologists (e.g. Star et al. 2018) suggest that the quest for walrus ivory may have been one reason for the initial settlement of Greenland by the Norse – perhaps just as Thule Inuit had moved eastwards across the Arctic in search of meteoric iron. Indeed, between the twelfth and fourteenth centuries, the Greenland trade of walrus ivory may have held a near monopoly in western Europe (Star et al. ibid.), and was an obviously major aspect of Norse productive activities. As such, it was an early form of more commercial resource extraction from Greenland's marine environment to meet the demands of wider markets.

Over time, however, the heavy reliance of the Norse colonies on long-distance trade networks as well as their dependence on a few trade items to sustain them would have made apparent the inherent social and economic vulnerabilities that tend to be characteristic of frontier regions – or places that sit on the edge of regions, far removed from the centres of power and decision-making. The Greenlandic settlements also relied on ships from Norway to bring vital supplies and to take away trade items, but the maritime link was difficult and often intermittent across the rough seas of the North Atlantic (Vésteinsson et al. ibid.). The living resources that the Norse settlements extracted from Greenland and the commodities they produced for export were also subject to the fickle nature of the market and changing consumer taste. Arneborg (ibid.) argues that the declining demand for walrus products in western European markets in the Late Middle Ages – partly because the softer ivory from African elephant tusks was becoming more widely available – may have been one of the factors that triggered the depopulation and eventual abandonment of the Greenland settlements. And as McGovern (1985) points out, coloured enamel had replaced walrus ivory as an ecclesiastical and secular decorative item by the fifteenth century. The cooling climate, however, also meant an increase in drift ice between Greenland and Iceland, which made it far more difficult for the trade ships that came from Norway to reach the Norse districts in Greenland. Increasingly cut off and isolated, with an eroding and diminishing trade network and less demand in Scandinavia and Europe for Greenlandic-produced commodities, life in the farms and villages of the farthest western regions of the Norse world would have been become more precarious and the rest of the Nordic world would likely have felt far more distant.

Later European interest in Greenland was mercantile or knowledge about it emerged as a consequence of other Arctic exploratory concerns. Early English explorers such as Martin Frobisher and John Davis led voyages in the late sixteenth century, with Robert Bylot and William Baffin sailing further north in 1615 and 1616, that attempted to find a northern sea passage to access Asian silks, spices, and other trade goods through the waters of the North American Arctic (Baffin had previously acted as pilot on James Hall's fourth voyage to Greenland in 1612 in search of silver and a way to the Northwest Passage). Sailing near the coasts of Greenland, they met with groups of Inuit and Davis produced the first written description of West Greenlanders. During the voyages of Frobisher and Davis, however, skirmishes occurred with the Inuit on Baffin Island and in Greenland, with some Inuit being killed by the English mariners and a few of Frobisher's crew going missing.

During this early period of searching for a northwest passage to access the riches of Asia, Greenland itself was not viewed as a place for the extraction of resources. However, based on an assay of a piece of black stone he brought back to England from his first expedition to Baffin Island in 1576, Frobisher believed he had discovered gold. Frobisher and his principal commercial backer Michael Lok established the Cathay Company and secured financial support and investments from Britain's merchant and royal classes, including Queen Elizabeth I, and two mining expeditions to Baffin Island set out in 1577 and 1578 (Ehrenreich 1998). Both ventures were ambitious – the first consisted of three ships, 145 crew, and eight miners, but the second was made up of 15 ships with over 400 men, 147 of whom were miners. As Ehrenreich discusses, these were militaristic expeditions and the base for the mining operations exhibited characteristics of a colonial extractive outpost in an area where Inuit lived, hunted, and fished.

Frobisher's expeditions involved the digging, loading, and transport of ore, and they represented the first prolonged contact between Inuit and Europeans on Baffin Island. The activities of the mining expeditions had an impact on the Inuit – there was conflict between them and the sailors and miners, and a number of Inuit were killed or captured, while the journals that recorded the expeditions described the Indigenous inhabitants as savage, devious, and murderous (Ehrenreich ibid.). Frobisher brought back what he thought was more than 1,000 tons of gold, but assayers in England declared the ore from both voyages to be worthless rock. This did not inspire further mining ventures for a while, but following the descriptions of whales, walrus, and narwhals in the accounts and journals of Davis and Baffin, European whalers were enticed to Greenland's waters for another form of resource extraction. This marked the beginning of commercial voyages to Greenland and the eastern part of what is now the Canadian Arctic (Williams 2009). And it was European whaling that had the first significant and more enduring and deeper impact on Greenland's Indigenous population, as well as on the country's marine resources. Competitive whaling between the English and the Dutch for the Greenland right whale – known as the bowhead in the western North American Arctic, European whalers thought of it as the right whale to catch, as it did not sink when harpooned and killed – in the waters around the Svalbard archipelago in the early seventeenth century resulted in the depletion of whale populations. This then led to the seeking out and exploitation of new whaling grounds, resulting in an international dispute over rights to what were considered the open northern seas. Increasing competition for whales led to a European rediscovery of Greenland and interest in its resource potential, but this also coincided with Denmark asserting its historical claim to sovereignty over the northern North Atlantic.

Danish colonisation: enclosing Greenland as a resource territory

In 1721, Hans Egede, a Norwegian-Danish Lutheran priest from Bergen on Norway's southwest coast, was appointed royal missionary and dispatched on a voyage to Greenland's west coast by King Frederik IV of Denmark and Norway. The king was concerned that the Norse settlements had been left so isolated from Scandinavia over the previous three centuries that their inhabitants had either remained Catholic or had lapsed into heathen beliefs and practices and were in need of being re-Christianised – this time as Lutherans, given that the Reformation had not reached Greenland's shores. Egede's voyage marked the reassertion of Danish interest in Greenland but it was driven by an imperative that was distinct from other European colonial ventures, including Denmark's own activities overseen by the Danish West India Guinea Company in what became the Danish Antilles or Danish Virgin Islands. In the West Indies, Danish colonies and plantations had been established from the mid-seventeenth century for the explicit purpose of the production and economic exploitation of resources, but at the time when Egede's mission was being planned, Greenland was still considered to be an integral part of the Nordic world under the joint Danish-Norwegian crown. Indeed, Greenland was thought of as both a Scandinavian inheritance and a dependency, not a place to lay claim to, colonise, and settle. It was a place to rediscover and the desire was to reconnect its people, who were thought to be isolated and adrift, with the wider Scandinavian homeland.

Attempts had been made in the fifteenth and sixteenth centuries to revive contact with the Norse colonies, but it was three expeditions dispatched by

Christian IV between 1605 and 1607 that 'rediscovered' Greenland and aroused great excitement in Denmark about its forgotten possession and the possibility for renewed trade with the Norse Greenlandic communities and the exploitation of marine resources as well as the potential for great mineral riches being found there (Etting 2009). When the first expedition captured two Inuit and took them back to Copenhagen, the king described them as Danish subjects. More Greenlanders were brought to Denmark on subsequent voyages, and following Christian IV's death in 1648 (he had been monarch since 1588), his son the new king Frederik III instructed that more expeditions should be sent out. Frederik III is said to have added a polar bear to the arms of the Danish monarchy in 1666 to symbolise Greenland's place within the realm.

Christian IV had granted a trade monopoly in Greenland to a group of Copenhagen merchants in 1636 following the foundation of the Grønlandske Kompagni a year earlier, although the shipowners seemed to be interested only in whaling rather than in any possibility of mining being a lucrative enterprise (Etting ibid.). Frederik IV – the son of Christian V and crowned King of Denmark and Norway in 1699 – was keen to re-establish trade links with the Norse Greenlandic communities, and this was another reason for Egede's expedition. With the support of the king and Bergen merchants, Egede established the Bergen Greenland Company (which came to be more usually known as the Bergen Company) to establish and administer Danish-Norwegian colonies and organise trade in Greenland.

Egede, who left for Greenland with his wife Gertrud Rask and their sons Poul and Niels, found no trace of living Norse settlers when they arrived at the outer edge of the Western Settlement in July 1721, although considerable evidence of their farms and dwellings was to be found around the coasts of the inner Nuuk and Ameralik fjord systems. However, he did encounter Inuit who were living around the headlands, islands, and outer skerries of what is now known as Nuup Kangerlua (the Nuuk Fjord region). They practiced forms of mobility and seasonal activities that covered an extensive marine and land resource base and their social and economic system was characterised by sharing and exchange. They were also embedded in networks of exchange with other Inuit groups that involved travelling great distances by kayak and umiaq along the west coast. These networks were notable for allowing the exchange of different foods, and Inuit trade also centred around other animal products such as furs and baleen. They also traded minerals and stone such as iron, soapstone, and flint that were important for hunting and fishing implements as well as for making household items. Some of this exchange occurred during summer events called *aasiviit* (singular *aasivik*, meaning 'summer place').

Aasiviit were also great social occasions – kinship ties and close social associations were celebrated, forged, and reaffirmed, marriages were arranged, stories were told, and knowledge about animals and places was exchanged (Marquardt and Caulfield 1996). Egede turned his attention to establishing a mission and trade station at Habets Ø near Kangeq and began to convert the Inuit living in the area or those who passed through to Christianity. This marked the beginning of more than 230 years of Danish colonial rule over Greenland's Inuit – although initially Greenland was more accurately part of the Dano-Norwegian realm until it was ceded entirely to Denmark in 1814 following the transfer of Norway to Sweden. In 1728, with the financial aid of the Bergen Company, Egede relocated his base across the mouth of the fjord and established Godthåb at the site of present-day Nuuk. The Bergen Company ended up having little success with trade in Greenland, but its activities did mean a reassertion of sovereignty and Egede's trade and mission station established Godthåb as the administrative centre for Danish colonial rule. The foundations of Egede's original house site can still be seen on Habets Ø.

Hoping to establish a viable and lucrative trade network based on marine mammal products (mainly seal and whale oil, and sealskins, fish, and fox furs), the Danish authorities assumed responsibility for trade in 1726. Under King Christian VI, trading rights in Greenland were transferred to state-sanctioned independent companies following the demise of the Bergen Company – first to Danish merchant Jacob Severin in 1733 (for whom Jakobshavn, sometimes then known as Jacobshavn, and now known as Ilulissat was named with the establishment of a trade station there in 1741) and then to the General Trade Company (*Det almindelige Handelskompagni*), which began operating in 1747. It was given the task of administering all settlements and trade until 1774, when the Danish government formed the Kongelige Grønlandske Handelskompagni or KGH (Royal Greenland

Figure 1.1 Kolonihavn – the colony harbour in Nuuk – where the KGH established its main base of operations. Hans Egede's statue is on the small hill in the background (Photograph Mark Nuttall).

Trade Company, in English). The establishment of the KGH initiated a state-run trade monopoly in Greenland based on the exploitation of marine and terrestrial living resources that was to last from 1776 until reforms were introduced in 1950.

During the eighteenth century, there was a progressive expansion of Danish activity along the north and south of Greenland's west coast as trade and administrative centres and settlements known as colonies, stations, and factories were established by Severin, the General Trade Company, and later by the KGH, such as Sukkertoppen (Maniitsoq), Hosteinsborg (Sisimiut), Egedesminde (Aasiaat), Jakobshavn (Ilulissat), Upernavik, and Julianehåb (Qaqortoq). As well as the colonies, smaller trading posts (*udsteder* in Danish, literally 'outposts' or 'out places') were established by the KGH to serve the smaller Inuit settlements and hunting camps.

Many *udsteder* were experiments for establishing regional centres for resource exploitation, as the KGH extended its activities and the commodification of Arctic wildlife over greater stretches of the west coast. Some *udsteder* were abandoned after a few years, while others gradually grew into larger villages and became smaller trade hubs themselves. Kangersuatsiaq in the Upernavik district along the northwest coast, for example, was established in 1800 as Prøven (meaning 'the test') to try a new technique for hunting beluga with nets and soon grew into a profitable settlement for the KGH. The Danish trading stations that were set up along the coast in the eighteenth and nineteenth centuries, were designed to cement and consolidate Copenhagen's authority over Greenland and the extraction of its living marine and terrestrial resources, but also to act as a deterrent to competition from others, such as the Dutch whalers and traders who were active and had been operating in North Atlantic waters for at least a hundred years before the Egede mission and who, along with English and Scottish whalers, were becoming more interested in the potential of working in Greenlandic waters. The Danes were eager to regain their political and economic position and advantage over access to and control over their Arctic resource space, dominate trade, assert sovereignty, and prevent other European traders from interfering with whaling and sealing in Greenland.

Even if they were engaged in a colonial enterprise that was not so different in its aims and objectives to other imperial powers, including other fur trade activities such as those organised and administered by the Hudson's Bay Company from 1670 in what is now known as Canada, the Danes continued to act in Greenland according to a belief that they were expanding trade in an inherited territory that was part of their Nordic realm and that they had rights to do so. By way of contrast, for example, the activities of Spanish and Portuguese colonialists in the Americas rested on papal authority and the 1494 Treaty of Tordesillas. This granted them rights to explore, discover, colonise, and occupy new territories, and in the process, erase Indigenous populations that may be in the way of them doing so. The Danish colonisation of Greenland differed in that the use of military force, active Indigenous resistance, large-scale conflict, or massacre did not characterise it, even if Inuit were drawn into the service of the trade, came to reside more in fixed settlement patterns, and were converted to Christianity, which undermined Indigenous social structures and spiritual beliefs. As mentioned above, they were already considered Danish subjects by the time Egede's trade and mission activities had started.

The Danes were not looking to eradicate Inuit hunting practices and economies along Greenland's coast, but they sought to maintain, expand, and profit from them. Skilled hunters were needed to provide the products of the trade. Inuit were converted to Lutheranism and young Greenlandic men were trained as catechists (often being sent to Copenhagen as part of their theological education) to assist in its spread, but there was no interest on the part of the Danes in a form of colonialism that implemented policies of social, cultural, and economic improvement. All efforts went into developing the trade by establishing stations and exploiting Inuit productive activities along the west coast. Customary hunting areas became resource bases that were essential to the KGH endeavour.

Hunting and fishing in the service of the trade had to be encouraged and sustained. For this to happen, the Danes argued that Inuit needed to be protected from too many colonising and civilising forces that altered their culture and the hunting way of life – even if, ironically, Danish activities affected how Indigenous forms of trade and barter were practiced (Gad 1982: 25). This resulted in a gradual territorial enclosure over much of what was then known as Greenland's geographical extent – but it was not a process of the privatisation of land and property rights and the dispossession of people and their removal from their traditional resource base. Rather, rights to trade were protected by the Danish monopoly and vested in the KGH. In the process, Greenland gradually assumed the shape of one enclosed – and closed – territory. Inuit moved within and around customary hunting places, continuing to follow a seasonal round, or settled and sought new opportunities at the places where the organisation of trade was centred and from where it was being expanded.

That said, this territorial enclosure still exhibited and deployed what are recognised as processes that allow the capitalist formation of space and resources (cf. Sevilla-Buitrago 2015) – a transformation of the non-commodified surroundings in which Inuit hunted and fished into the resource bases organised around the activities of the colonies, whaling stations, and factories. These resource bases represented a form of enclaving and the consolidation of distinct hunting territories (with the commodification of marine and land animals as well as various species of fish) that were overseen by regional trade officials. Significantly, Greenland was not a settler colonial society – Danes, Norwegians, and other Europeans living and working there were priests, catechists, and KGH employees, not people who moved to take advantage of land and homesteading. Indeed, it was not possible to go to Greenland and work unless it was in the service of the church and the trade. There was no intended dissolution of Indigenous society, eradication of culture, or annihilation of the Inuit population, but this was not to say they were not subjected to a process of colonialism that set in motion the creation of a new society and the biopolitical making of Greenland as territory and nation (Gad 1973).

By 1782, most Inuit who lived along the west coast, from Cape Farewell in the south to Upernavik in the north, had come into permanent contact with Danes and other Europeans at trade stations and religious missions. The previous year, the KGH had divided Greenland into northern and southern districts, each with an inspector to supervise trade and oversee general administration. It continued to be the case that Danes and Norwegians living and working in Greenland were restricted mainly to missionaries, colonial inspectors, traders, and their assistants. The trader in each colony and small settlement occupied a position of authority and power over the Inuit that was based on his responsibility for collecting and distributing trade goods and for paying wages in kind. Typically, the trader and other KGH employees who came to work in the settlements seldom stayed for more than a few years before returning home, but some did stay for decades, married Greenlandic women with whom they had children, and learned to speak Greenlandic fluently.

Unlike colonial governments extending their reach over other parts of the world and its peoples at the time, Danish policy towards Greenland was isolationist and based on a sense of paternalism, even if there was no desire to implement social reforms for the benefit of the people of the kingdom's northern territory. Again, this had the expressed intention of maintaining hunting and fishing and protecting Inuit culture in the interest of the trade. Hunting and fishing had to provide the basic needs of the Inuit population, but it also had to sustain trade activities and provide the seal blubber, skins, whale oil, baleen, dried fish, and other commodities that were extracted and made by Greenlandic efforts and labour and sent by ship each year to Copenhagen. Yet, in the early decades of trade activities, Greenland's population grew and gradually became one of mixed Greenlandic and Danish/Norwegian descent which challenged colonial governance in the service of the trade (Priebe 2017; Seiding 2011).

The KGH published its Order of 1782 that year, which set out a way to manage and cultivate Greenlandic society through strict rules for the economic, practical, and moral conduct of KGH employees in their relations and interactions with Greenlanders. By this time, a stratified society with several socio-economic groups was taking shape in Greenland. The Order identified five groups: Trade and whaling employees; mission employees; the West Greenlandic population of Inuit origin (who were also considered to be 'real' Greenlanders); *blandinger*, literally 'mixtures' or 'mixed-bloods,' who were the children of European men and Greenlandic women; and foreigners, who were usually other Scandinavian and European employees of the KGH (Gad 1982: 28–32; Rud 2017). Although the *blandinger* had become a large group by the 1780s, marriages between Europeans and those who were defined as 'real' Greenlanders were now banned by the Order (Gad ibid.). This was done to protect the trade – as what became to be defined as the 'real' or 'genuine,' traditional Greenlandic family (for which we should read 'unmixed') was considered a vital social unit that provided the social and cultural context for raising and training boys as hunters and girls as hunters' wives.

The Order laid down rules that Greenlanders were not to be 'spoiled' by foreign goods, but a Danish-speaking Indigenous Greenlandic élite, whose members had Danish/Norwegian and Greenlandic family backgrounds and social, cultural, and economic connections to Denmark, emerged and continued to reproduce itself socially through education, intermarriage, and occupation (Rasmussen 1986:

144–145). In a Foucauldian sense, the activities of the KGH, institutionalised by the Order of 1782, employed instruments of biopower that served to discipline Greenlanders, creating 'docile bodies' at sites and enclaves such as mission stations, churches, and trade colonies, and through the idea of the 'real' and 'genuine' Greenlander as noble hunter. It enacted a biopolitics that administered, managed, and controlled Greenlandic lives and bodies so that life and the activities essential to the trade and the production of resources and commodities could be optimised effectively (cf. Foucault 2007).

By the end of the eighteenth and beginning of the nineteenth centuries, then, the majority of Greenland Inuit along the west coast were involved in a trading economy that was based mainly on whaling as well as the procurement of seal oil, sealskins and fish, and other items such as fox furs, polar bear furs, and narwhal tusks. They had become dependent to a considerable extent on trade goods to supplement their diet and mode of production. New techniques for efficient and more productive hunting and fishing were introduced by the KGH, such as seal nets and the development of commercial hunts for certain species such as beluga whales, as in Prøven in Upernavik district where beluga were trapped in a narrow passage by setting nets as they migrated past the settlement. The hunting of large whales was organised from whaling stations, which were often run under the supervision of Greenlanders who were the sons of mixed marriages. Whaling

Figure 1.2 The KGH established trading stations along the west coast of Greenland during the eighteenth century. Upernavik was established in 1772 (Photograph Mark Nuttall).

attracted people to those stations, many of which were established on the central west coast, from different parts of Greenland, and the trade required a range of skills, so there were opportunities for Greenlanders to train as carpenters, coopers, and other craftsmen (Gad 1982).

By 1901, it was estimated that almost two-thirds of the population of North Greenland and almost one-third of the population of South Greenland were of mixed Danish (and to some extent, Norwegian) and Greenlandic descent (Priebe ibid.; Seiding ibid.). Danish painter Andreas Riis Carstensen, who was noted for his depictions of maritime scenes, was disappointed when he first visited Greenland in 1884 to find that the Greenlanders he met on the north central coast did not look like the traditional, exotic people he expected to encounter:

> There was nothing in the appearance of those people that showed them to belong to a race different from my own, only their language was unlike anything I had heard before. Some had dark hair, and others blonde.; most of them had beards; and if some of them, on a closer inspection, showed traces of Eskimo descent, there was not a physiognomy with pure Eskimo features amongst the whole.
>
> (Carstensen 1890: 14)

Despite the changes to Inuit society and culture that occurred as a result of Danish involvement in Greenland and the emergence of an upper social stratum, the majority of the Inuit population continued a hunting and fishing lifestyle, even if it was reshaped by certain technological interventions and economic purpose and their customary surroundings were reshaped as resource spaces essential to the trade. Indeed, as pointed out above, this was actively needed, encouraged, and promoted by the KGH to sustain its activities. While there were experiments in South Greenland with cattle rearing, sheep farming, eiderdown gathering, and shark liver oil production, seals and other marine mammals continued to provide the mainstay of the local economy, with skins and blubber underpinning the trade economy.

Nonetheless, the KGH was concerned that young Greenlanders were losing the vital hunting skills, including kayak building, that were needed to be a real hunter and that more people were being drawn to settle close to the trading posts, even if this was an inevitable outcome (and an initial intention) of the trade system it oversaw. The KGH encouraged a dispersal of the population, especially in the sealing districts, and continued to nurture and perpetuate the idea and image of the hunter as the 'real,' 'genuine' Greenlander living a virtuous life. The KGH was keen more than ever to assert that the hunter embodied *the* Greenlandic way of life. The true Greenlander, it was implied – and asserted through the biopolitics of administering Indigenous lives and consolidating the categories of social stratification – was someone who hunted and practiced what was defined as a traditional hunting lifestyle on water, ice, and land, and was not influenced by Danish culture to the same extent other Greenlanders were in the larger colonies along the coast (Gad ibid.; Nuttall 1992).

While the trade was based on the procurement of products from living marine and terrestrial beings in what became well-defined resource spaces along the west coast, in the nineteenth century there was some interest in minerals and the development of mines. The practice of geological exploration and research in Greenland has a long history that has involved the mapping of territory, the geospatial assembling of resources, and a political discussion of the economic possibilities of extraction – something I return to in later chapters. Formal geological exploration and work on the assessment of Greenland's mineral potential date from the early part of the nineteenth century, with the first surveys carried out along the west coast by German geologist Karl Ludwig Giesecke, who collected rocks and minerals and looked for copper and ores between 1806 and 1813. Some local mining took place at various locations in Greenland for coal, such as the Nuussuaq peninsula and Disko Island in the northwest, to provide a source of fuel for the colonies and trading stations. From the 1850s, mining for cryolite at Ivittuut in South Greenland (which was used to produce soda in making glass and soap, as well as in aluminium production) contributed further to the extraction of Greenland's resources and its profitability for the Danish authorities and much of it was exported to the US. An attempt was also made to extract copper at Innatsiaq (Josva) in South Greenland in the 1850s, but the mine was eventually abandoned until in reopened in 1907. It then remained operational until 1914.

A turning point in Denmark's approach to colonisation came in the early 1830s. This followed the report of Danish naval officer Wilhelm August Graah's expedition to South and East Greenland. In March 1828, Graah had been sent from Denmark by Frederik IV to map the southeast coast of Greenland, look for evidence of Norse settlement there, and secure Danish sovereignty. Even though Hans Egede's travels on the west coast and subsequent Danish surveys in southern parts of Greenland during the eighteenth century had not revealed any evidence that there were living descendants of the Norse Greenlanders from the Western Settlement, the Danes had not given up hope of finding the descendants of the Eastern Settlement, or at least indications that they had colonised the east coast. Having spent a year in Greenland preparing for the expedition, Graah and his crew set off in March 1829 from Nanortalik, which was established in 1770 and was the most southern of the Danish colonies at the time (a permanent trading station was located there in 1797).

Graah met Inuit in Southeast Greenland, but concluded that not only were there no surviving descendants of the Norse Greenlanders to be located, there had never been any Norse settlements on the east coast (Graah 1837). His report undermined, in a sense, the Danish rationale for the rediscovery and colonisation of Greenland in the first place. With no Norse villages or farms to find, the Danes had to revise their approach to dealing with the social, economic, and welfare needs of Greenland's Indigenous inhabitants and the increasingly apparent societal divisions that were emerging as a result of the trade activities, social stratification, and occupational differentiation (Rud ibid.). Graah himself proposed some changes that would be focused on an improvement in living conditions, and he was appointed a member of the Greenland Commission when it was established in 1835 to address the question of the trade monopoly and the development of the school system and the health service (Rud ibid.).

The Greenland Commission can be seen as a first official attempt at designing a new form of Greenlandic society that would continue to provide the basis for the trade system by strengthening a sense of pride in the hunting culture, but also advance Greenland towards what the Danes considered to be a higher level of civilisation for Greenlanders (Rud ibid.). Graah argued for the promotion of cod fishing, the construction of houses that would replace traditional dwellings, and the expansion of the school system. He also advocated for sending young Green-landers to Denmark for education, and 12 travelled there between 1837 and 1843. The idea was that once they returned from Denmark, educated and 'civilised' in a Scandinavian way, they would be able to contribute to the further education of the rest of the Greenlandic population. Other reforms included the establishment of two teacher training colleges in Nuuk and Ilulissat, and those educated there strengthened the core of the Greenlandic élite, many of whom were considered *blandinger* (Rud ibid.).

Largely due to these reforms and the work of an educated élite, the idea of Greenland as a country, and of Greenlanders (at least those in the south and along the west coast) as a people with a distinct social and cultural identity and who shared a sense of history inhabiting an emerging, modernising nation, began to take shape during the nineteenth century. Key to this was the development of Kalaallisut as a written language and the establishment of a printing house in Nuuk during the 1850s. The newspaper *Atuagagdliutit* (meaning *distributed reading matter*) was first published there in 1861, initially appearing monthly. Its signifi-cance lay not only just in being a source of news in Greenlandic about Greenland and the outside world, but for providing a medium for cultural expression and political discourse. It paved the way for the beginning of a Greenlandic literary tradition, played a significant role in the spread of literacy along the coast, con-tributed to the idea of Greenland as a nation, and gave a voice to those who were arguing that Greenlanders should be involved in the administration and govern-ance of their own land and its resources.

Extending Danish sovereignty

Despite the reach of KGH trade activities and the shaping of regional resource spaces along the west coast, some Inuit populations remained relatively isolated, having little or no contact with Europeans. The northern and eastern parts of Greenland were imagined, represented, and approached as remote and empty. While the west coast was gradually drawn into the structures and institutions of Danish administration and the process of territorial enclosure enacted by the KGH trade machinery, and while the education and health systems were trans-forming the lives of many people, North and East Greenland were considered to be places apart from the rest of the country. It was only later in the nineteenth century and early decades of the twentieth that Danish exploration and geolog-ical mapping efforts were tasked to discover more of Greenland, to make these remoter areas known and legible, assess their resource potential, and support the assertion of sovereignty over the farthest reaches of the kingdom.

Inughuit were visited in Northwest Greenland by Scottish explorer John Ross and Anglo-Welsh explorer William Parry in 1818, who captained, respectively, the vessels *Isabella* and *Alexander* during their British Admiralty-organised expedition to discover a route to the Northwest Passage; and while the Indigenous people of the Avanersuaq region became used to explorers and whalers as regular summer visitors and trading partners during the nineteenth century (Vaughan 1991), the KGH thought the area too remote to extend trade activities there. British Royal Navy officer Douglas Clavering encountered a group of 12 Inuit in Northeast Greenland in 1823, while in command of HMS *Griper* on an expedition that was part of another Admiralty project – a survey to determine the shape of the earth (indeed, this is the only recorded contact between Inuit and Europeans in what today is an uninhabited part of Greenland, except for a small military detachment and a couple of scientific research and weather stations); and as mentioned above, Graah met with Inuit in Southeast Greenland in 1829–1830. In 1884, over 160 years since the first encounters between Greenlanders and Danish missionaries and traders on the west coast, Gustav Holm wintered with the people of Ammassalik (the present-day Tasiilaq region) on the east coast.

Inuit from the southeast coast had been travelling to, trading in, and settling around the Cape Farewell area in the most southerly part of Greenland during much of the nineteenth century. They brought stories with them of people living further north along the east coast (e.g. Gulløv 1995). Holm's expedition to Ammassalik was seen as leading to the 'discovery' of these isolated East Greenlanders and Danish administrators and ethnographers classified them as an untouched tribe with a pristine culture. This influenced Danish attitudes towards East Greenland, especially policy discussions about the area and its development. Holm was accompanied by Johannes Hansen (known as Hanserak) and Johann Petersen who came from the educated West Greenlandic élite – they later played a role in the colonisation of East Greenland and in the establishment of Ittoqqortoormiit. Hanserak was a catechist who kept a diary during the expedition, and which was published in a series of sketches and observations about the Ammassalimmiut (the people of Ammassalik; East Greenlanders also refer to themselves as Iivit in contrast to Kalaallit of the west coast) and their homeland in *Atuagadliutit*.

Before Holm's journey, Graah's own narrative of his expedition to Southeast Greenland contains descriptions of encounters with Inuit, but while he regarded those he met as pagan, he noted they possessed some items of European manufacture, including "some female ornaments, such as beads, bits of red ribbon, &c.," and observed people wearing "gay clothing of the latest fashion," leading him to conclude that they "are no strangers to our settlements" (Graah ibid.: 68). Indeed, Graah later described meeting with people who were on their way to barter "with their countrymen of the West coast, bear, seal, and dog skins for articles of European manufacture, and especially for spear and arrow heads, knives, needles, handkerchiefs, and tobacco" (ibid.: 81). However, he felt that the Inuit on this part of the east coast had not yet been changed and disrupted by civilisation. Visiting a family who cooked seal meat for the expedition members, Graah wrote that the "countenance" of two women:

> had nothing of the ordinary Greenland physiognomy. Their whole appearance, indeed, presented none of the usual characteristics of their race, and, in particular, they had neither the prominent belly, nor the corpulence of, their countrywomen on the west coast. They were, both of them, above the middle size, and were remarkable for their clear complexion, their regularity of feature, and the oval form of their heads. They were more cleanly, also, than their countrywomen of the West coast, who, indeed, trick out their persons on Sabbath-days with beads, and ribbons, and variegated seal-skins, but are filthily dirty all the rest of the week.
>
> (Graah ibid.: 70).

Even if Graah described meeting Inuit who he supposed had not previously seen a European (e.g. ibid.: 83), most people along the southeast coast were involved at the time – even tangentially – in trade networks in which European items circulated, or at least they had access to some of those items.

Danish missionary and linguist Christian Wilhelm Schultz-Lorentzen (1928: 14) remarked how the "discovery" of the Ammassalimmiut by Holm, further north than Graah had travelled, was exciting for Danish ethnologists at a time when they were looking for "the original foundation of the Greenlander" and for a group that had no previous contact with Europeans. In *The Ammassalik Eskimo* (1914), William Thalbitzer drew on the ethnographic studies that had emerged from the first expeditions as well as his own visit to the area. He wrote that

> In the first mentioned district (West Greenland) the mind and body of the Inuk (Eskimo) is so strongly Europeanized, that a considerable power of criticism is required to distinguish the true remnants of the original culture in modern implements and mental products. A valuable corrective to the results of my first journey was my stay in East Greenland, where the original culture was still easily detected.
>
> (Thalbitzer 1914: 326)

Holm's expedition of 1884–1885 was followed by other exploratory ventures to map East Greenland. These included the Amdrup Expedition (1898–1900) led by Danish naval officer Georg Carl Amdrup, the ill-fated Danmark Expedition (1906–1908) to the northeast coast led by Ludwig Mylius-Erichsen, and Knud Rasmussen's Fourth, Sixth, and Seventh Thule Expeditions between 1919 and 1933. The expeditions led by Amdrup and Rasmussen had central ethnographic components, including ethnological studies and the description and collection of items of material culture, while Mylius-Erichsen also wanted to gather further information about Inuit in the region who had only been visited by Clavering in 1823.

Representations of East Greenland as remote and extreme and a place that was geographically and culturally separated from West Greenland were perpetuated in exploration literature, which also emphasised its purity and unspoiled nature as an unmapped polar region (see Graah ibid.; Koch 1955; Mikkelsen and Sveistrup 1944; Nansen 1897). In this work, East Greenland is imagined as a masculine world of Danish Arctic adventure, a backdrop for heroism, sacrifice, and conquest (Koch ibid.: 8–14). At the same time, Danes and West Greenlanders, who read these narratives, as well as Hanserak's diary in *Atuagadliutit* thought of the East Greenlanders as heathen and backward, whose culture was characterised by shamanism, magic and murder, infanticide and infidelity. They felt their responsibility to the people on the east coast was to educate them and bring to them the benefits of Danish – and West Greenlandic – society and the trade economy.

There were no permanent, fixed year-round Inuit settlements in East Greenland when Holm and his colleagues visited in 1884. This was a nomadic space in which extended families moved seasonally within and around a wide area of fjords, islands, and headlands in search of hunting and fishing grounds. Several families came together to form a winter household, but individual family units lived in tents during the spring and summer. Holm saw this and reported his concerns that the Ammassalimmiut subsisted from meagre resources and would disappear unless there was Danish intervention. He argued that what he encountered was a loosely organised and nomadic people threatened by starvation, that blood feuds, infanticide, manslaughter, and suicide were rife, and that they were in need of protection (Mikkelsen and Sveistrup ibid.: 78).

In 1824, the German Moravian Brethren established a mission at Friedrichsthal (Frederiksdal in Danish and now known as Narsarmijit, it is the most southerly settlement in Greenland) in the Cape Farewell area, and this attracted Inuit from the southeast coast throughout the nineteenth century to trade and settle (Gulløv ibid.; Jensen 2004). After Holm's visit to Ammassalik, people in the area also began to move south to gain access to European trade items as well as to benefit from being in the orbit of the Moravian mission activity. The Danish administration decided to establish a trading post at Ammassalik in 1894 – Johan Petersen became the colony manager – and also step up efforts to convert the East Greenlanders to Lutheranism. The trading post and the accompanying elements of the colonial state assemblage meant that the Ammassalimmiut were brought into the territorial enclosure of Danish jurisdiction, but it also prevented more immigration into South Greenland and the further conversion of East Greenlanders to Christianity by the Moravians (Gulløv ibid.).

At the same time, there were tensions between Denmark and Norway over the sovereignty of much of the east coast, especially the more northerly stretches, to which Norway had long held claim. Although Norway became independent when its personal union with Sweden was dissolved in 1905, it had kept its own constitution, Storting (parliament), and institutions following the Convention of Moss of 1814, when Charles XIII of Sweden became king of Norway. Norway continued to assert its own sovereignty and ideas about territorial expansion in the North Atlantic. In doing so, it only recognised Danish claims to the west coast with its colonies and settlements. For Denmark, then, the establishment of the trading station at Ammassalik was also an act of asserting Danish control over territory as well as settling the East Greenlandic population. Norwegian hunting, trapping, and fishing activities had taken place periodically on the east coast since 1889.

In a sense, Norway considered Greenland's east coast to be an extension of Arctic territory that stretched from Svalbard and across the Greenland Sea, and over which it felt it had a claim because of prior occupancy. From time to time, Norway made a claim to Greenland through the Norse settlements – it was argued that the Icelandic migrants to Greenland in the tenth century (including Eirík the Red) had originally mostly been Norwegians. Norwegian trappers also hunted Arctic foxes and polar bears on Jan Mayen between 1900 and 1920, and Norway was given jurisdiction over the island by the League of Nations in 1921. The Norwegian government considered Northeast Greenland to be uncolonised, a region that Denmark had shown no interest in. As such, for Norway, it was *terra nullius*. However, in July 1919, Norway's foreign minister Nils Claus Ihlen made a verbal declaration (known as the Ihlen Declaration) to Denmark that Norway would present no difficulty in respect of Danish sovereignty over the whole of Greenland. Questions were raised in the Norwegian parliament, however, as to whether this was binding on Norway and it continued to occupy parts of Northeast Greenland through the presence and activities of the hunters and trappers there.

In July 1923, the Norwegians invited the Danes to enter into negotiations on the question of East Greenland. An agreement in 1924 – to last twenty years – allowed Norwegians to use unpopulated areas on the east coast as long as they did not interfere with the livelihoods of Inuit in Ammassalik. Norwegian ships and their crews were allowed access to the east coast and were given the right to land, to hunt, and fish, and to overwinter. Norway was also given the right to erect meteorological, telegraphic, and telephonic stations. It had already established a meteorological station on Jan Mayen in 1921, and so was becoming dominant in the skills and techniques of northern North Atlantic weather observation as well as having a certain power over the knowledge it gathered. This agreement, however, did not change the respective Danish and Norwegian attitudes: Denmark contended that it possessed sovereignty over the whole of Greenland, while Norway continued to assert that not all of Greenland had been occupied and brought under Danish administration and that the uncolonised areas were still terra nullius. In fact, Norway argued that if these parts of East Greenland ceased to be terra nullius, then they would be under Norwegian sovereignty because of continued Norwegian use. Trapping and meteorology became key to the assertion of Norwegian territorial rights and claims.

The increased activity by Norwegian hunters and trappers in Northeast Greenland came to be seen as a threat to Danish sovereignty and to Denmark's policy of protecting Greenlanders from outside influence (notwithstanding, that is, their incorporation into the trade economy and the effects of 200 years of Danish intervention and intrusion into their bodies, minds, lives, and lands). The Danish government passed two laws in April 1925, one on fishing, hunting, and navigation in Greenlandic waters, the other on the administration of Greenland. The first reserved the right to hunt and fish in Greenlandic waters for Danish subjects (including Inuit) who were settled in Greenland and for those who were able to qualify for special licences. The second law divided Greenland into three administrative provinces – North, South, and East – and reserved all commercial activities in Greenland to the Danish state. Norway objected to this, arguing that it could not apply to areas, especially on the east coast, where Denmark had not demonstrated colonisation, occupation, and sovereignty (for an excellent and comprehensive history of Norwegian and Danish trapping in Northeast Greenland during the twentieth century, see Mikkelsen 2008).

To extend sovereignty over East Greenland, famed Danish explorer Ejnar Mikkelsen rallied private support for his idea of establishing a Greenlandic settlement at Ittoqqortoormiit (Scoresbysund), nearly 1,000 kilometres northeast of Ammassalik, in order to force Norwegian hunters and trappers to move out of the area. Initially, the Danish government did not endorse the venture, but funds were raised from among the Danish public. A committee that included Gustav Holm as a member was formed to oversee the establishment of Ittoqqortoormiit as a planned community. There were concerns over population increase in the Ammassalik area and worries that people there were concentrated in a few locations that could not support their needs as a resource base. The establishment of Ittoqqortoormiit would not only ensure Denmark could extend its administrative reach over the east coast and further assert sovereignty there, it would allow for the formation of an ideal hunting community that would protect and nurture what was viewed as a true, real Greenlandic culture, something that administrators and ethnographers felt had been disappearing on the west coast and was now again threatened around Ammassalik. Mikkelsen argued that Danes and Greenlanders shared parental responsibilities for East Greenlanders, who were described as "children of the moment" (Mikkelsen and Sveistrup ibid.: 230).

Mikkelsen sailed to Ittoqqortoormiit in summer 1924 with a group of construction workers and scientists to set up the community. Archaeological evidence indicated that Inuit had been living in Jameson Land and coastal and inland areas further north. This – together with the evidence of Inuit occupation and resource exploitation of Northeast Greenland from Clavering's encounter a century before – lent further support to the venture. It also bolstered the argument that Danish subjects had lived all along the east coast, exploring and living along its extensive coastline and utilising resources there and so, given this historic occupancy, Norway could not consider it terra nullius. Even if it could not be established that Norse had settled on the east coast, Inuit had certainly travelled and lived along its entire length, it was claimed. And the fact that Graah had found no traces or evidence of Norse settlement on the east coast was used by Denmark to any possibility that Norway could assert that Norwegians had colonised East Greenland from Iceland. The new community's location was selected in an area of previous Inuit settlement identified by archaeologists. In this way, Mikkelsen sought to re-establish Inuit occupancy and resource use. Some sites of historic Inuit dwellings were even dug out and used as the foundations for new houses. Scientists mapped out the surrounding area and its local resource potential.

Ammassalimmiut were promised better hunting opportunities if they were to relocate to Ittoqqortoormiit, and some 80 people from several extended families were selected to move. Seventy of these people were Iivit and they were accompanied by several West Greenlandic officials and their families, including Johan Petersen, the former colonial administrator in Ammassalik on the journey north in 1925. A number of older people died during the first winter in their new settlement, mainly from diseases they had contracted during a stopover in Ísafjörður in Iceland. In establishing the new community of Ittoqqortoormiit, the settlers had to make and shape a new resource space through hunting and fishing and through daily engagement with the non-human entities in the surroundings that were to become their homeland. The dispute over the legal status of East Greenland was settled in September 1933, when the Permanent Court of International Justice in The Hague accepted that Denmark exercised sovereignty over the whole of Greenland – geological mapping, the assessment of resource potential, and archaeological evidence or previous resource use by Inuit were key to the Danish case.

Much of the data Koch had gathered during years of geological research was central to the Danish argument (Cavell 2008; Hudson 1933; Preuss 1932). With this decision, and the development of Ittoqqortoormiit and the re-making of an Indigenous identity in relation to a place where there was no living memory of dwelling in, the Danish enclosure of Greenland and its construction as a bounded territory was complete. Yet, as historian (and head of the Greenland Representation in Copenhagen) Jens Heinrich (2018: 29) points out, this consolidation of Danish administration was done without the involvement of Greenlanders other than in how they were used for the diplomatic battle for international recognition of Danish sovereignty over Greenland. The Greenlandic councils did not participate in any major foreign policy decisions. Greenlandic society was still considered fragile and Greenlanders were believed to be in need of protection. Foreign relations were still under the control of Denmark, but as I will discuss in Chapter 4, World War II and the Cold War period brought Greenland into a new strategic role in the world. And it was following the end of World War II that Denmark also began transforming Greenland society and economy. This provided the context for the emergence of a move to self-government, the realisation of self-determination, and a strong articulation of aspirations for independence.

References

Arneborg, Jette. 2003. "Norse Greenland: Reflections on settlement and depopulation." In James H. Barrett (ed.) *Contact, Continuity and Collapse: the Norse colonization of the North Atlantic.* Turnhout: Brepols Publishers, pp. 163–181.

Arneborg, Jette. 2012. "Norse Greenland dietary economy ca. AD 980- ca. AD 1450. Greenland Isotope Project: Diet in Norse Greenland AD 1000- AD 1450." *Journal of the North Atlantic,* Special Volume 3: 1–39.

Arneborg, Jette. 2018. *Greenland: Approaches to historical Norse archaeology.* Copenhagen: National Museum of Denmark.

Arenborg, Jette. 2021. "Early European and Greenlandic walrus hunting: motivations, techniques and practices." In Xénia Keighley, Morten Tange Olsen, Peter Jordan and Sean Desjardins (eds.) *The Atlantic Walrus: Multidisciplinary insights into human-animal interactions.* London: Academic Press, pp. 149–167.

Behrisch, Erika. 2003. "'Far as the eye can reach': Scientific exploration and explorers' poetry in the Arctic, 1832–1852." *Victorian Poetry* 41 (1): 73–92.

Carstensen, A. Riis. 1890. *Two Summers in Greenland: An artist's adventures among ice and islands, in fjords and mountains.* London: Chapman and Hall.

Cavell, Janice. 2008. "Historical evidence and the Eastern Greenland case." *Arctic* 61 (4): 433–441.

Craciun, Adriana. 2011. "Writing the disaster: Franklin and *Frankenstein*." *Nineteenth-Century Literature* 65 (4): 433–480.

Dodds, Klaus, and Mark Nuttall. 2018. "Materialising Greenland within a critical Arctic geopolitics." In Kristian Søby Kristensen and Jon Rahbek-Clemmensen (eds.) *Greenland and the International Relations of a Changing Arctic: Postcolonial paradiplomacy between high and low politics.* London and New York: Routledge, pp. 139–154.

Dugmore, Andrew J, Christian Keller and Thomas H. McGovern. 2007. "Norse Greenland settlement: reflections on climate change, trade, and the contrasting fates of human settlements in the North Atlantic islands." *Arctic Anthropology* 44 (1): 12–36

Ehrenreich, Robert. M. 1998. "Mining, colonialism and culture contact: European miners and the indigenous population in the sixteenth-century Arctic." In A. Bernard Knapp, Vincent C. Pigott and Eugenia W. Herbert (eds.) *Social Approaches to and Industrial Past: The archaeology and anthropology of mining.* London and New York: Routledge, pp. 109–119.

Etting, Vivian. 2009. "The rediscovery of Greenland during the reign of Christian IV." *Journal of the North Atlantic*, Special Volume 2: 151–160

Fisher, Wayne W. 1954. "The passing of Godthaab." *Foreign Service Journal* 31 (3): 30–31, 52.

Fogelson, Nancy. 1985. "The tip of the iceberg: The United States and international rivalry for the Arctic, 1900–25." *Diplomatic History* 9 (2): 131–148.

Foucault, Michel. 2007. *Security, Territory, Population.* New York: Palgrave MacMillan.

Franks, Jill. 2006. *Islands and the Modernists: The allure of isolation in art, literature and science.* Jefferson, NC and London: McFarland & Company.

Friesen, Max. 2022. "Ancestral landscapes: Archaeology and long-term Inuit history." In Pamela Stern (ed.) *The Inuit World.* London and New York: Routledge, pp. 17–33.

Gad, Finn. 1973. *The History of Greenland II: 1700 to 1782.* Montreal: McGill-Queen's University Press.

Gad, Finn. 1982. *The History of Greenland III: 1782–1808.* Montreal: McGill-Queen's University Press.

Graah, Wilhelm August. 1837. *Narrative of an Expedition to the East Coast of Greenland, Sent by Order of the King of Denmark, in Search of the Lost Colonies.* London: J. W. Parker.

Grønnow, Bjarne and Jens Fog Jensen. 2003. *The Northernmost Ruins of the Globe: Eigil Knuth's archaeological investigations in Peary Land and adjacent areas of High Arctic Greenland.* Meddelelser on Grønland – Man and Society 29. Copenhagen: Danish Polar Center.

Gulløv, Hans Christian. 1995. "'Olden times' in South Greenland: New archaeological investigations and the oral tradition." *Études/Inuit/Studies* 19 (1): 3–36.

Hanrahan, Maura. 2017. "Enduring polar explorers' Arctic imaginaries and the promotion of neoliberalism and colonialism in modern Greenland." *Polar Geography* 40 (2): 102–120.

Hayes, Isaac Israel. 1871. *The Land of Desolation: A personal narrative of adventure in Greenland.* London: Sampson Low, Marston, Low and Searle.

Heinrich, Jens. 2018. "Independence through international affairs: How foreign relations shaped Greenlandic identity before 1979." In Kristian Søby Kristensen and Jon Rahbek-Clemmensen (eds.) *Greenland and the International Relations of a Changing*

Arctic: Postcolonial paradiplomacy between high and low politics. London and New York: Routledge, pp. 28–37.

Hudson, Manley O. 1933. "An important judgment of the world court." *American Bar Association Journal* 19 (7): 423–425.

Jensen, Einar Lund. 2004. "*Uiarnerit*: A historical study of immigration from East to West Greenland in the nineteenth century." *Études/Inuit/Studies* 26 (2): 23–46.

Koch, Lauge. 1955. *Report on the Expeditions to East Greenland 1926–1939, conducted by Lauge Koch*. Part II. *Meddelelser om Grønland* 143 (2). København: C.A. Reitzel.

Kroon, Aart and Bjarne Holm Jakobsen. 2010. "Coastal environments around Thule settlements in Northeast Greenland." *Geografisk Tidskrift* 110 (2): 143–154.

Madsen, Christian K. 2019. "Marine shielings in medieval Norse Greenland." *Arctic Anthropology* 56 (1): 119–159.

Markham, Clements R. 1853. *Franklin's Footsteps*. London: Chapman and Hall.

Marquardt, Ole and Richard A. Caulfield. 1996. "Development of West Greenlandic markets for country foods since the 18th century." *Arctic* 49 (2): 107–119.

Mason, Owen K. 2020. "The Thule migrations as an analog for the early peopling of the Americas: Evaluating scenarios of overkill, trade, climate forcing, and scalar stress." *Paelo-America* 6 (4): 308–356.

McGhee, Robert. 1984. "Contact between native North Americans and the Medieval Norse: A review of the evidence." *American Antiquity* 49 (1): 4–26.

McGhee, Robert. 2000. "Radio carbon dating and the timing of the Thule migration." In M. Appelt, M. Berglund, and H.C. Gullov (eds.) *Identities and Cultural Contacts in the Arctic: Proceedings from a conference at the Danish National Museum*. Copenhagen: Danish National Museum, pp. 181–191.

McGhee, Robert. 2009. "When and why did the Thule move to the Eastern Arctic?" In Herbert Maschner, Owen Mason and Robert McGhee (eds.) *The Northern World AD 900–1400*. Salt Lake City, UT: University of Utah Press.

McGovern, Thomas H. 1985. "The Arctic frontier of Norse Greenland." In Thomas H. McGovern (ed.) *The Archaeology of Frontiers and Boundaries*. New York: University of New York Academic Press, pp. 275–323.

McGovern, Thomas H. 2017. "Zooarchaeology of the Scandinavian settlements in Iceland and Greenland: diverging pathways." In A. Umberto, H. Russ, K. Vickers, & S. Viner-Daniels (eds.) *The Oxford Handbook of Zooarchaeology*. Oxford: Oxford University Press, pp. 1–23.

Mikkelsen, Ejnar and P. P. Sveistrup. 1944. *The East Greenlanders Possibilities of Existence, Their Production and Consumption. Meddelelser om Grønland* 134(2). København: C.A. Reitzel.

Møbjerg, Tine. 1986. "A contribution to Paleo-Eskimo archaeology in Greenland." *Arctic Anthropology* 23 (1–2): 19–56.

Nuttall, Mark. 1992. *Arctic Homeland: Kinship, community and development in Northwest Greenland*. Toronto: University of Toronto Press.

Nansen, Fridtjof. 1897. *Farthest North*. London: Archibald Constable and Company, 2 vols.

O'Reilly, Bernard. 1818. *Greenland, the Adjacent Seas, and the North-West Passage to the Pacific Ocean, Illustrated in a Voyage to Davis's Strait during the Summer of 1817*. London: Baldwin, Craddock and Joy.

Priebe, Janina. 2017. *Greenland's Future: Narratives of natural resource development in the 1900s until the 1960s*. Umeå: Umeå Universitet.

Preuss, Lawrence. 1932. "The dispute between Denmark and Norway over the sovereignty of East Greenland." *American Journal of International Law* 26 (3): 469–487.

Rasmussen, Hans-Erik. 1986. "Some aspects of the reproduction of the West Greenlandic upper social stratum, 1750–1950." *Arctic Anthropology* 23 (1–2): 137–150.

Roucek, Joseph. 1951. "The geopolitics of Greenland." *Journal of Geography* 50 (1951): 239–246.

Rud Søren. 2017. *Colonialism in Greenland: Tradition, governance and legacy*. London: Palgrave.

Schledermann, Peter. 2000. "Ellesmere: Vikings in the Far North." In W. W. Fitzhugh and E. I. Ward (eds.) *The North Atlantic Saga* (pp. 248–256). Washington DC: Smithsonian Institute, pp. 248–256.

Schultz-Lorentzen, C. W. 1928. "Intellectual culture of the Greenlanders." In M. Vahl, G.C. Amdrup, L. Bobé and AD.S. Jensen (eds.) *Greenland: The colonization of Greenland and its history until 1929*. Copenhagen: C.A. Reitzel.

Seaver, Kirsten. 1996. *The Frozen Echo: Greenland and the exploration of North America, ca. AD 1000–1500*. Stanford, CA: Stanford University Press.

Seiding, Inge. 2011. "Intermarriage in colonial Greenland 1750–1850: Governing across the colonial divide." In Michelle Daveluy, Francis Lévesque and Jenanne Ferguson (eds.) *Humanizing Security in the Arctic*. Edmonton: CCI Press, pp. 111–126.

Sevilla-Buitrago, Alvaro. 2015. "Capitalist formations of enclosure: Space and the extinction of the commons." *Antipode* 47 (4): 999–1020.

Seward, A.C. 1922. *A Summer in Greenland*. Cambridge: Cambridge University Press.

Star, Bastiaan, James H. Barrett, Agata T. Gondek and Sanne Boessenkool. 2018. "Ancient DNA reveals the chronology of walrus ivory trade from Norse Greenland." *Proceedings of the Royal Society B*. 285 (1884), http://doi.org/10.1098/rspb.2018.0978.

Sutherland, Patricia. 2000. "The Norse and Native North Americans." In W. W. Fitzhugh and E. I. Ward (eds.) *The North Atlantic Saga*. Washington DC: Smithsonian Institute, pp. 238–247.

Thalbitzer, William. 1914. "The Ammassalik Eskimo: Contributions to the ethnology of the East Greenland Natives." *Meddelelser om Grønland* 39: 1–755.

Vaughan, Richard. 1991. *Northwest Greenland: A history*. Orono: The University of Maine Press.

Vésteinsson, Orri, Claus Andreasen, Inge Bisgaard, Kenneth Høegh, Birger Lilja Kristoffersen, Anja Jochimsen, Paul Ledger, Pia Lynge, Christian Koch Madsen, Mikkel Myrup, Georg Nyegaard and Henning Sørensen. 2016. *Kujataa - a subarctic farming landscape in Greenland: A nomination to UNESCO's World Heritage List*. Kujataa Municipality: Greenland National Museum and Archives.

Williams, Glyn. 2009. *Arctic Labyrinth: The quest for the Northwest Passage*. Toronto: Viking Canada.

Williamson, Geoffrey. 1953. *Changing Greenland*. Sidgwick and Jackson Limited.

2 Transformation and design

Following World War II, Denmark ended the isolationist policy towards Greenland that had formed the basis for colonial governance and trade since the eighteenth century and began a process of modernisation and economic restructuring (in Chapter 4, I discuss how Greenland became a *de facto* American protectorate during the war and the implications this had for Danish administration afterwards as well as Greenland's geo-strategic positioning in world affairs). The United Nation's Charter of 1945 – and its emphasis on self-determination as a right of all people – was an initial impetus for Danish decolonisation policy. The Greenland Commission (*Grønlandskommissionen*), which was an outcome of a meeting between the national councils of Greenland and the Danish Prime Minister Hans Hedtoft in Godthåb, was set up by the Danish government in 1948. It was tasked with examining what were thought of as the defining problems of Greenland in terms of social, economic, political, cultural, and administrative development. Nine sub-commissions were established, involving Danish civil servants and other experts as well as representatives from Greenland. It led to the production of a report in 1950 known as G-50, which articulated a political and economic plan to design and build a modern Greenlandic society – more about which I discuss in this chapter.

The G-50 report scoped out ways for the development of infrastructure and for the improvement of health and living conditions, as well as the development of commercial fisheries. Initially, the idea was for the state to invest in infrastructure and the social and health systems, with private investment encouraged for the fisheries (which was never realised, so Denmark invested in the building of factories in the towns on the west coast). Through the establishment of two provincial councils and 62 communal councils in 1911, Denmark had set in process a motion whereby Greenlanders had a greater institutional voice in its relations with the Danish government, but could also act to strengthen Greenlandic society (Heinrich 2018). The two provincial councils were brought together in 1950 in a union that formed the Greenland Provincial Council (*Grønlands Landsråd*). Effectively, this assembly acted and functioned as a provincial government until Home Rule was introduced in 1979 and the Greenlandic government and parliament were established.

Rather than seeing themselves as responsible for continuing to preserve what they considered to be an original and traditional form of Greenlandic society

DOI: 10.4324/9781003175421-3

and culture that was based on hunting, Danish colonial authorities and administrators now turned from ways of governing that had been informed by ideas of protecting hunting families living in small and relatively isolated settlements and exploiting specific resource spaces and, by doing so, maintaining the social, cultural, and economic foundations that made the trade system possible. Yet, Janina Priebe (2017) argues that the modernisation and industrialisation of Greenland along the lines of state-led development initiatives from the 1950s was not an entirely new policy approach. She shows how these trends had been discernible since 1900 and were part of a colonial policy that even then was attempting to respond and adjust to a changing world. As she points out:

> Since Greenland's colonization by Danish-Norwegian traders and missionaries in the 18th century, natural resources were at the center of debates about the country's future. Yet, it was not before around 1900 that administrators, scientists, and private entrepreneurs alike brought forward comprehensive visions of a new dimension that challenged the traditional colonial approach of isolating Greenland's hunting culture. Natural resources industries, imagined as large-scale ventures, were at the core of their ideas of how to govern the dependency in the future.
>
> (Priebe ibid.: 2)

Priebe (2015) also suggests that the more formalised and official idea of Greenland as a resource frontier began to take shape around 1905, as private entrepreneurs imagined the future of the territory's economy and formed a consortium to establish industrial fisheries, scope out the potential for mines and the exploitation of new resource spaces, and encourage new forms of business and enterprise.

From around this time, the KGH was no longer responsible for much of the administration and governance of Greenland and the Danish Ministry of Interior gradually began to assume control. It was the consortium's view that Denmark's trade monopoly had become economically unsustainable and, with its focus on protecting the Greenlandic population and the territory's resources from foreign intervention, was unable to respond to the rather dramatic environmental change that was being experienced at the beginning of the twentieth century. This was particularly evident in the southern parts of the country, where the sea was warming and affecting marine mammal hunting. As a result, there was a shift in the ecosystem and a fluctuation in seal populations, and Hamilton, Lyster, and Otterstad (2000) show how climate change began to influence – indirectly – a major transition in resource use from seal hunting to commercial cod fishing as a new form of living marine resource extraction.

Other initiatives for economic ventures that could replace seal hunting were also being tried in the first couple of decades of the twentieth century. One of these was sheep farming and the cultivation of an agricultural frontier and its potential as an expansive resource space. Animal husbandry was not an unknown practice in southern Greenland, but its economic prospects had never been thought to be especially promising by the KGH. The Norse had brought cattle, sheep, and goats from Iceland in the tenth century; cattle and sheep were sent from Denmark to Habets Ø in 1723 at the request of Hans Egede; cattle, sheep, pigs, goats, chickens, geese, and ducks were kept at most KGH colonies when they were established along the west coast in the eighteenth century; and Moravian missionaries also kept flocks of sheep at their mission stations, which also had small gardens to produce vegetables (Gad 1973: 402; 1982: 198).

In 1780, Norwegian trader Anders Olsen, who had arrived in Greenland in 1742 and had worked for Jacob Severin's trade company and established the trade colonies of Julianehåb and Sukkertoppen (Maniitsoq), began to raise cattle and sheep at Igaliku in South Greenland (which was known as Garðar in Norse Greenland, and had been prominent as the seat of the Greenlandic bishop). Olsen married a Greenlandic woman by the name of Tupernat and he also established Igaliku in 1783 as a regional trade station. Their son Johannes Andersen took over the farm on Olsen's death in 1786. Gad (1982: 200) reports that Andersen and his descendants were the only people in Greenland to practice sheep farming as well as cattle rearing as an occupation. Andersen sold mutton and beef and he kept a garden and grew vegetables, but he received no support from the colonial administration for his farming venture. Although the farm succeeded, he nonetheless depended on hunting seals and fishing for subsistence.

Around this time, concern was expressed by colonial administrators that other Greenlanders who wanted to begin farming would subvert the trade in South Greenland and encourage those it depended on for producing the very commodities that sustained the KGH to abandon hunting and fishing. However, some missionaries and traders felt that many of those they labelled as 'half-Greenlanders' were idle and slovenly and would soon become a burden to the country and harm the trade, and that farming could provide a way for them to earn a living. Andersen was held up as an example of a diligent and hard-working 'half-Greenlander.' The KGH, however, did not consider it possible that animal husbandry could be maintained on a larger scale, given Greenland's soil and climate, and that others could follow Andersen's example with any chance of success, so farming was not promoted (Gad ibid.). At the beginning of the twentieth century, however, the problem of the declining seal catch and concern over the impoverishment of much of South Greenland's population led to a reconsideration of its possibilities.

In 1906, Greenlandic priest Jens Chemnitz had brought 11 sheep from the Faroe Islands to establish a farm at Narsarmijit – the site of the former Moravian mission station at Frederiksdal (Sørensen 2007: 27). This interested those reformers in the colonial government who suggested that animal husbandry, farming, and agriculture could be ways of modernising the economy of South Greenland as well as dealing with the problem of the declining seal catch, which they now deemed unproductive as a trade activity. Inspired by ideas and hopes of what an agricultural frontier promised, the KGH established sheep breeding stations in South Greenland between 1913 and 1915 to provide sheep and training in animal husbandry to Greenlanders who were interested in turning from seal hunting to sheep farming (Sørensen ibid.). In 1915, Lindeman Walsøe imported 170 sheep from Iceland for the breeding station at Julianehåb and these were lent to people to help them start their own flocks (Hayashi 2013). Sheep farming proved to be an effective initiative – and it is no coincidence that the modern farms were often established on the fertile land where the Norse Greenlanders had their farms and settlements (Rose et al. 1984: 65–67). In 1924, Otto Frederiksen, who is widely recognised as being one of the pioneers of sheep farming and who adopted it as a full-time occupation, established a farm on a site that had been occupied by Eirík the Red – and this and many other pioneering farms are still productive. Sheep farming soon became the third largest industry in Greenland (Sørensen ibid.).

Colonial status was superseded in 1953 when Greenland became an integral part – officially a county – of the Kingdom of Denmark. In Chapter 4, I discuss how the process that led to this policy shift had come about following the end of World War II. Anne Kristine Hermann (2021) suggests, however, that Denmark's motive in incorporating Greenland into the Kingdom of Denmark in this way was driven by a desire to protect its territorial interests in the Arctic and North Atlantic. The UN Charter was, she writes, a threat to the integrity and geographical extent of the kingdom. By granting Greenland the status of a county and thereby according it a status as an integral part of the kingdom, Denmark ensured that the territory would be removed from the UN's list of colonies entitled to independence. She argues that Greenland became, in a sense, 'a camouflaged colony' that continued to be subject to Danish jurisdiction and planning.

The ending of colonial rule marked the beginning of an era that was to be characterised by far-reaching economic transformation, infrastructural transfiguration, and profound social and cultural upheaval. Key to the implementation of all this was the establishment in June 1950 of the Greenland Technical Organisation (*Grønlands Tekniske Organisation* [GTO]), which assumed considerable authority over executing the Danish policies of development outlined in the G-50 plan. Also, by 1950, the KGH no longer asserted its monopoly over trade. There were social problems to overcome and solve and new economic enterprises to develop, but Greenland was in some ways also considered by the GTO to be a technical and engineering challenge. Before World War II, building activity in Greenland was restricted to dwellings, warehouses, stores, hospitals, and schools. Wood was the main material with concrete used occasionally for foundations. From the 1950s, the GTO began constructing standardised – and greatly subsidised – housing in towns and settlements with a greater use of concrete. The GTO was placed within the framework of the Ministry of Greenland in 1955, when it formed the Construction Committee (*Anlægsudvalget*), which was comprised of civil servants and representatives from construction and property development companies. Housing was also required for the increasing numbers of Danes who were moving to Greenland to work in the many roles needed to implement the G-50 policy, but a better quality of accommodation provided to them, along with higher wages and other workplace privileges, highlighted some of the inequalities of modernisation and development.

Ten years after the G-50 plan was first implemented, the Minister for Greenland set up the Greenland Committee (*Grønlandsudvalget*) of 1960 to review the development process and assess the prospects for the further reconstruction of

Greenland. The committee's report, known as G-60, set out another ten-year plan for the development of Greenland from 1966 to 1975. The G-60 committee proposed the formation of the Greenland Council (*Grønlandsrådet*) to co-ordinate the planning work in Greenland. The Greenland Council was established in 1964 – it had five political representatives from Greenland and five from Denmark and was dissolved on the introduction of Home Rule in 1979. G-60 outlined a plan that would concentrate the fishing industry, economic development, and provision of education, health, and social care services in four towns – Godthåb (Nuuk), Frederikshåb (Paamiut), (Holsteinsborg (Sisimiut), and Sukkertoppen (Maniitsoq) – which were in the ice-free zone on the west coast, and so this allowed for the construction of open water harbours (Nuttall 1992, 1994).

Policies for centralisation and urbanisation were implemented according to the G-60 plan. Small, often remote settlements, which were once key to the KGH-controlled trade, were now deemed as unproductive and unprofitable by the Danish authorities (especially those in what were considered 'outlying' sealing districts, where the population had been dispersed by the KGH in the nineteenth century), so many were closed and people were moved from these communities to live in apartment buildings in the growing west coast towns. To create a new urban environment and a new economy, many of the settlements that were once vital to the trade enterprise and which needed to be maintained and sustained for it by extracting living marine and terrestrial resources now needed to be erased. This disinvestment in the settlements led to a process of unbecoming along many parts of the coast and a process of renewal and transformation in a few towns (cf. Fraser 2018).

Inuit Pinngortitarlu

For six years, between 2013 and 2019, my colleague Lene Kielsen Holm from the Greenland Climate Research Centre and I co-ordinated a collaborative project in partnership with a group of Nuuk residents who had been born in settlements in Nuup Kangerlua that were closed as part of the G-60 process. We called this project 'Inuit Pinngortitarlu,' which loosely translates as 'People and Environment.' *Pinngortitaq* is often translated from Greenlandic as 'environment' or 'nature,' but I shall discuss further in Chapter 6 how this has a richer meaning in Greenlandic – essentially referring to the Indigenous view of seeing the world as always coming into being and always taking shape. Lene was originally from South Greenland and she passed away in January 2021 – a tremendous loss to her family, friends, and colleagues, but also to Indigenous-led research in Greenland and the Arctic more generally. A champion of community-based research, Lene's work was always guided by a collaborative approach to the production of knowledge, and she had participated in a number of projects that contributed greatly to our understanding of climate change, human-environment relations, and Indigenous use of resources. One of her most notable works is *The Meaning of Ice* (Gearheard et al. 2013), a beautiful book co-edited with scholars and Indigenous community experts in northern Alaska, Canada's Nunavut Territory, and Greenland.

Inuit Pinngortitarlu was a partnership Lene and I developed together in that spirit. Like other research projects we worked on at the Greenland Climate Research Centre, we were guided by the important principle of anchoring it in the community. Its origin lay in our interest to understand the effects of climate change in the context of other pressing processes of change in the Nuup Kangerlua area, such as the legacies of the social and economic transformations people had experienced since the 1950s, the political transformation of Greenland, and the presence of extractive industry. We were also concerned at the time by the absence of any real understanding of – or desire to recognise and acknowledge – the relations and entanglements between humans, non-humans, and the wider environment in the social and environmental impact assessments produced for a number of mining projects that were being scoped out, including those for a major iron ore mine at Isukasia, some 150 kilometres northeast of Nuuk (Nuttall 2012, 2017). We recognised that people with whom we worked and talked with about extractive industries and their plans for developing mines in Nuup Kangerlua were frustrated that company executives, mining engineers, and consultants would often describe the area as an empty space and a wilderness. Our work contributed to an understanding of the region – already established by Indigenous knowledge, archaeology, and cultural history – as a place in which people have long engaged in a complexity of rich and intricate social relations with the environment with animals and other living and non-living entities.

I was privileged to work with Lene as we sketched out the project, did the fieldwork and organised workshops, carried out interviews, and participated in dialogue sessions with community members, civil society groups, and NGOs. By researching local knowledge and perceptions of weather, climate, and environment, the use of living marine and terrestrial resources, the growing importance of tourism and leisure, and the political, social, and environmental aspects of extractive industries that await in the future, Inuit Pinngortitarlu mapped the past and present use of Nuup Kangerlua and its outer coastal areas and speculated on what may affect the region in the coming decades. Part of this work involved analysis of historical records, hunting and fishing catch lists, log books and diaries, contemporary accounts, place names and stories, and mapping the hunting and travel routes of people who have lived in the inner and outer parts of Nuup Kangerlua. We were interested not only in historical occupancy but also in contemporary movement and use, including the many plans for mines in the area and the geological prospecting that accompanied them. A further aspect of this project placed more recent changes in archaeological and historical context, and one essential part of it, a PhD dissertation project, focused on the impact of environmental change on historic Inuit and Norse cultures, and on adaptation strategies in relation to changes in sea ice, climate, and the environment (Lennert 2017).

Inuit Pinngortitarlu involved numerous Nuuk residents – and we invited them to participate in a variety of ways, such as contributing family stories about hunting, fishing, animals, travel, and campsites. In this way, we sought to understand movement, mobility, use, and occupancy of Nuup Kangerlua as a set of diverse configurations and patterns of human-environment relations and multispecies engagement. Crucially, however, it was centred around a core of four local experts, the research partners with whom we developed the scope and direction of the project: Angunnguaq Josefsen, a hunter and fisherman, whose family originated from Uummannaq, a small island in the inner fjord, and who grew up in Kapisillit, a village some 75 kilometres northeast of Nuuk, deep in the fjord; Vittus Nielsen, a fisher and wildlife officer, who originated from Qoornoq, a community in the central part of the fjord that was closed in 1972; Kaaleeraq Tobiassen, who was born in Kangeq and was a former trade manager with the KGH (and who also spent some time in Savissivik in Qimusseriarsuaq/Melville Bay); and Marius Tobiassen, Kaaleeraq's brother, a hunter and fisher, also from Kangeq, and who once worked on trawlers in Greenland's offshore waters. All four were born between the late 1930s and late 1940s and had hunted and fished from childhood. Marius and Angunnguaq have passed away in the last couple of years.

During the project, Lene and I travelled on a number of occasions with Vittus, Kaaleeraq, and Marius to their former home villages as well as to other settlements that were abandoned in the inner fjord and along the outer stretches of the coast, and with Angunnguaq to Kapisillit and the places where his parents and grandparents had lived. We spoke about the places where they, their families, and their ancestors had hunted, fished, and trapped, and where they had camped. They told us about what these surroundings and the non-human entities that comprise them meant to people and to themselves, and we heard and recorded their stories of use and movement in the past. They told us what the environment and weather was like when they were growing up, what animals were important to them in different seasons, and how – in the case of Vittus, Kaaleeraq, and Marius – they felt when they and their families were moved to Nuuk. For them and many others, having their home settlements closed and being rehoused in apartment buildings was not just a feeling of dislocation but of rupture, a feeling of having one's self, body, and identity severed from the surroundings that nurtured them. Our Inuit Pinngortitarlu archive is a vital record of human-environment relations, much of which Lene and I had been working on together to translate into English. The narratives of all four partners, which were recorded and written in Kalaallisut, constitute a profound account of a life in small communities that was centred on hunting and fishing before the planned phase of urbanisation and centralisation ended a pattern of seasonal movement. There is still much to transcribe, translate, and write.

It would be wrong to think of this – and even to romanticise it – as a traditional life of hunting and fishing that was not unchanged by Danish intervention, however. As I illustrated in Chapter 1, many of Greenland's fixed villages and permanent settlements were established in the eighteenth and nineteenth centuries as religious missions or as trade stations by the KGH – some were themselves experiments in new ways of procuring resources for the trade economy – but they were placed in the traditional hunting and fishing territories of nomadic Inuit who had histories of extensive occupancy, often reaching back to Paleo-Inuit times, and who, despite the commodification of animals and fish, continued to depend on living marine and terrestrial beings for their survival. This is evident in Nuup Kangerlua (e.g. Gulløv 1997).

Kaaleeraq and Marius, for example, contributed knowledge and oral histories of life in Kangeq, which is on the same island where Hans Egede established his mission and trade colony in 1721 and where he first encountered Inuit when he arrived in Greenland. Kangeq means a promontory or headland and is situated at the mouth of Nuup Kangerlua, about 18 kilometres southwest of Nuuk. It is a rocky elevation that juts into the sea, forming a small bay which was used as a natural harbour. The area has a long history of occupancy. It attracted people from various parts of the coast because of its rich resource base and ice-free surrounding water and was an important winter settlement (Gulløv ibid.). It was the site of a German Moravian mission from 1754 until 1900, when the Moravians left Greenland (Gulløv ibid.); but Kangeq had become an official hunting and fishing station for the KGH in 1854 and reached its highest population of 155 in 1960. People also moved seasonally or permanently to other settlements in Nuup Kangerlua. Kangeq was closed down by the Danish authorities in 1974 and the 80 or so people who were living there at the time were moved to Nuuk.

Although most houses were abandoned and lie in ruins today, as do other community buildings such as stores and warehouses, Marius maintained his family home and returned to Kangeq whenever he could – often with Kaaleeraq – in spring, summer, and autumn to hunt and fish. Lene and I were able to join them on several visits to the old settlement and we also travelled with them around the low-lying islands and skerries that make up this part of the coast. One day in

Figure 2.1 Kangeq, which is situated at the mouth of Nuup Kangerlua, was closed by the Danish authorities in 1974 (Photograph Mark Nuttall).

September 2014, we sat in the living room of Marius's house and Kaaleeraq spoke about key hunting and fishing places. "Remember," he said, looking at a map of the area we had set out on the table, "these have been used as hunting places since time immemorial. There have been archaeological excavations in the area for some years, which shows that the area of Kangeq has been inhabited for more than 4,000 years." We were able to visit some of these ancient sites with Kaaleeraq and Marius, places on headlands and skerries filled with tent rings, food depots, the foundations of turf houses, and graves.

Kaaleeraq recounted part of the seasonal round when his grandfather was a young hunter at the end of the nineteenth century and the beginning of the twentieth:

> I was born on 15th January 1939, in Kangeq. My father was born on 19th August 1904, also in Kangeq. My grandfather was born in 1876. I think, according to the stories told to us, that he began to take part in hunting trips around 1895. They would go to Attorsuit, a place where many hooded seals were caught, and also where many other animals were taken, in the period of March and the end of April, even also in May. On their return from Attorsuit they rested in Kangeq, and around the middle of May they went to Qaquk. There they fished and hunted for lumpsucker, halibut, beluga and harp seals. In those days there were plenty of beluga. The belugas came into the area on their way to the north after overwintering in the fjords. The hunters would lead them into Qaquk until the tide was at its lowest and then catch all of them. They could collect and prepare lots of food – dried beluga meat, dried harp seal meat, and also dried halibut. In those days, the halibut grew big and there were plenty of them. When the hunters rested in Kangeq they would go to Kitsissut, to Napparutileeqqap Sallersua, to hunt for the fatty harp seals of the autumn, with which they would enlarge their winter supplies. As our late mother told us, they would go there in the month of August. The rest of the year, until early spring they would spend in Kangeq, from where they would go hunting for seabirds and seals.

Kaaleeraq, Marius, Vittus, and Angunnguaq recorded their accounts of how people – including themselves – continued this seasonal round of movement around their localities and their engagement with the environment and non-human until the transformations that were brought about with urbanisation, the development of the large-scale commercial fishing industry, and the closure of the villages in the 1960s and 1970s. While we travelled with Kaaleeraq and Marius to places they hunted and fished, we also walked with them on old paths along which hunters would carry their kayaks to the water's edge – to Qaguk, for instance, where they too hunted beluga, other whales, and seals and where they fished for lumpsucker; to Qaattorfik, a place for netting harp seals; to Illunnguit Kangilliit, where they would stay overnight when hunting seals and birds; and to the many places where they would prepare meat and fish for winter storage. Vittus had kept finely detailed records of his travel routes and the hunting and fishing places that were important to the Qoornormiut between the 1940s and 1960s. These records include maps, diaries, notebooks, and descriptions of places, such as Ilorsua, where a vital fishery for redfish was conducted; Nuugaarsuk, where food for the winter was produced; Innarsuaq, where various edible plants and angelica (kuannit) were collected; and Kuussuaq, where reindeer were hunted and Arctic char fishing camps would be set up.

Every place is named and multi-layered, and rich stories are associated with them, as is the case elsewhere along the coasts of Greenland where people have lived and still make their lives today (Nuttall 1992). These stories are vital records, constituting knowledge about use and occupancy as well as community events, non-human beings, and extraordinary happenings. Angunnguaq described to us how the inner part of the fjord was important for seal hunting, for fishing cod, gathering berries, and reindeer hunting. His family originated from Uummannaq, which was another site of a Moravian mission station in the nineteenth century. The Moravians closed their station at the beginning of 1900 and Angunnguaq's ancestors moved to a place called Qassinnguit, a little further east in the fjord and close to the remains of a Norse settlement. Having moved there from their homes on Uummannaq, they built and lived in stone and turf houses, and Angunnguaq remembered his father telling him how they used stone and other material from the Norse house and barn ruins to construct their homes. His family's house was placed high on a slope, an important location for keeping an eye out for beluga entering and leaving the fjord on their annual migration. Angunnguaq's family left Qassinnguit around 1930 and moved to Kapisillit, although they returned annually for many years, especially in summer and autumn, and used their former home as a seasonal hunting and fishing base.

In August and September 2013, we travelled with Angunnguaq to the sites of these former familial homes and to Itivi, a place where his family had a summer camp. This and other areas such as Ipiutaq, a little further in the fjord, were sites where people from different places in Nuup Kangerlua gathered to hunt and fish. They followed the seasonal movement of animals, but their procurement activities were also influenced by the demands of the KGH. By October, hunters and fishers could rely on ice forming on the sea. In the 1950s, Angunnguaq would go with his father to Taseraarsuk ('the place with big lakes') in mid-February. They would check on their fox traps along the coast and it was an important winter hunting area for catching seals by setting nets under the ice and fishing for cod, char, halibut, and redfish.

This seasonal round, based on people's dependence on marine mammals, fish, and other animals such as reindeer and Arctic fox – and influenced to a considerable extent too by the demands of the KGH – was practiced in similar ways along Greenland's west, north, south, and east coasts. Hunting and fishing may have been key to the activities and fortunes of the KGH, but they were activities that underpinned Inuit culture, provided food for family and household, and formed the basis for social organisation and for community practices of the sharing and distribution of meat and fish. Life was attuned to the migrations and presence of animals and fish, and where people lived and spent time at different periods of the year was a characteristic of seasonal variation. By the late 1960s and early 1970s, Greenlandic society along much of the west coast – even up to Upernavik to some extent – had been transformed. For many people, the livelihoods based on hunting and fishing as described by Kaaleeraq, Marius, Vittus, and Angunnguaq, in the Nuuk region – Nuup Kangerlua and along the outer coast – ended.

The emergence of political parties

Despite these changes, people remember and speak about some good things about life in Nuuk and other towns. There was improved health care, which aided population growth and development policy focused on a commercial fishing industry, and an export-oriented economy, which provided jobs and provided people with a regular wage. But, ultimately, the aim of this new social and cultural policy was to turn Greenlanders into Danish citizens, the Danish language was privileged over Greenlandic, and a number of Greenlandic children were separated from their families and sent to school in Denmark and placed with foster families as a social experiment in assimilation and modernisation (Bryld 1998; Thiesen 2023). As a friend of mine I have known for twenty-five years and who was born in Nuuk in the late 1950s put it to me once, "Our minds and bodies were colonised."

The number of Danes posted to Greenland to work as administrators, teachers, doctors, nurses, police officers, managers in the fishing industry and other businesses, technicians, and construction workers increased significantly, and so the history of post-colonial Greenland is also one of the mobility of people from outside the territory but still within the Danish realm, who moved there to take advantage of employment opportunities and even to experience a few months or a couple of years of Arctic adventure. Initially, some Danes such as carpenters, electricians, plumbers, and other tradespeople went to work in Greenland on short-term seasonal contracts for the summer and lived in temporary housing. Others, such as administrators and teachers, arrived on one- or two-year contracts, but some began to settle permanently. Danish teachers, for example, who were given postings in smaller villages in the central and north-western parts of the coast as well as on the east coast, kept dog teams and hunted and fished.

When I first did fieldwork in villages in North, South, and East Greenland between the late 1980s and mid-1990s, it was common to meet school teachers from Denmark who had been in Greenland since the late 1960s and early 1970s, sometimes having lived in several communities, and who hunted seals and caught cod and Arctic char for some of their family meals. Some spoke Greenlandic, although many did not (villages tended – as some still are – to be predominantly Greenlandic-speaking and not all residents knew or felt comfortable speaking Danish). Some were married to Greenlanders, although many were not. They were often respected for the techniques and skills they learned from Inuit, such as how to build a kayak or a dog sledge, and for their abilities to hunt and to process meat and prepare sealskins. I would visit them in their homes, and they would also visit households to drink coffee when there was a birthday celebration, for instance, or on other community social occasions, but in many cases, there was a noticeable social distance during everyday life between them and the people whose children they taught.

People would explain to me though that they would tend to feel *ajukkunneq*, which means a sense of inferiority, in the presence of Danish teachers. They said they would often feel this in the presence of other Danes too who were in positions of leadership or who had a powerful role, such as a doctor (Nuttall ibid.). The teacher – the *ilinniartitsisoq* (plural *ilinniartitsissut*) – was still, as local residents would point out, a *Qallunaaq*, a Dane, and, from the 1950s up to the introduction of Home Rule in 1979, considered part of the Danish workforce that was carrying out the modernisation and urbanisation of Greenland as mandated by the Danish state. Even in the late 1980s, and as I also experienced in North, South, and East Greenland during the first half of the 1990s, feelings of *ajukkunneq* were difficult for many people to dispel. Such feelings were also expressions of how people experienced themselves to still be positioned with a colonial social space.

At this time, it was the case that there were still few Kalaallit *ilinniartitsissut* living and working in the settlements, and while the Danish *ilinniartitsisoq* played a role as a key figure, engaging in daily life through the running of the school and the implementation of the curriculum, they would likely return home to Denmark for a few weeks each summer. In some ways, the Danish teacher in a settlement embodied many of the characteristics of the stranger, as Georg Simmel postulated. For Simmel, the stranger is someone who comes today and stays tomorrow – someone who is both near and remote, mobile, close to the community, yet an outsider (Simmel 1950). Simmel also saw the stranger as being a member of a system, but not strongly attached to it – this could not be said of the Danish *ilinniartitsisoq*, however, whose role as a teacher was to enforce a new system of education that was essential to the Danification of Greenland. This system remains in place, despite self-government, as Home Rule was a process of devolution by which responsibility for the institutions of governance, society, and economy were transferred from Denmark and emplaced in Nuuk. Self-Rule continues this process, which is not one of the Indigenisation of institutions and systems (as, for example, has been the intent in Nunavut in Canada). Education in Greenland – even at the level of the university in Nuuk – is part of the wider Danish system.

The programmes of social and economic change that were implemented in the 1960s and 1970s came with a strong technocratic impulse from Copenhagen, symbolised by the continued powerful role of the GTO, which was charged with building infrastructure and more housing. Gunnar Rosendahl, who was born in Paamiut in 1919, was GTO director from 1965 to 1989 and presided over this technical transformation. In the mid-1970s, one of the GTO's feasibility studies, in collaboration with the then Geological Survey of Greenland (GGU), was a systematic mapping of hydroelectric power potential in Southwest Greenland. At the time, it was anticipated that a mining boom would soon happen and the mining industry would be a major consumer of energy during exploration, construction, and exploitation phases – the GTO and GGU assessment was that the amount of electricity that could be produced through hydropower projects would far exceed the demand for public consumption, so there would be enough to supply the needs of international companies. By the late 1970s, however, the GTO did not consider it a public responsibility to provide such energy and so concentrated its efforts into further planning for hydropower to supply Greenlandic towns. Before that, however, the GTO's main tasks were in the construction of housing schemes. The GTO was reorganised as Nuna-Tek in 1987 when it became an agency under the Greenland Home Rule administration, and Rosendahl continued in his role for two years after it had been renamed. Nuna-Tek was closed down in January 1990 and its tasks disbursed to several independent companies.

Population growth in the 1960s was most notable in the towns as a result of movement and displacement from the settlements, and this placed demand on the GTO to provide more housing. Large apartment blocks – identified with letters or numbers – were built quickly. In what is now the centre of Nuuk, these buildings came to dominate the newly forming urban landscape as it spread quickly over the surrounding fells and hills. Grydehøj (2014) points out that, at the time, such residences – with electricity, heating, and running water – were celebrated by Danish administrators for the way they represented major improvements in living standards over 'traditional' Greenlandic housing, such as the turf and stone houses many people lived in, especially in the settlements. The largest of these new buildings was Blok P. Completed in 1966, Blok P was 64 flats long and five storeys high. When its construction was completed, it was the largest apartment block in Denmark and was regarded not just as a marker of successful development policy in Greenland, but as a symbol of modernisation and of the drawing in of the Greenlandic population into the mainstream of Danish life (Grydehøj ibid.).

I would argue that Blok P and other apartment buildings constructed in Nuuk and other towns along the west coast in the 1960s and 1970s represented what Adam Kaasa (2020) calls 'domestic monumentality.' In developing this idea, Kaasa refers to a major housing project in Mexico City called Nonoalco-Tlatelolco that was built between 1949 and 1964. Monumental in scale, it comprised 102 buildings that could house 100,000 people, but it was only a portion of what had been planned. Kaasa's notion of domestic monumentality has relevance for understanding the planning process for Greenland, especially in how G-60 was enacted. Nonoalco-Tlatelolco held, he argues, a narrative promise for the Mexican nation state, yet it was responsive to the domestic. In a similar way, the Danish urban intervention in Greenland's towns – the former trade colonies and administrative centres of KGH activities – and the construction of new housing was a signifier of the formal incorporation of Greenland into the Danish state as county, territory, and constituent part, but in moving families from settlements into towns and new neighbourhoods, the emphasis was on apartments as family units and the creation of new domestic spaces for Inuit and Danes in a modernising Greenland. These new buildings too held a narrative promise for the future.

As Grydehøj discusses, Nuuk's rapid growth was dependent on imported Danish designs as well as the materials, technologies, the policies, and the labour needed to build the new metropolis. Today, these apartment blocks – all of which were constructed rapidly and often maintained poorly – are in a state of dilapidation and badly affected by damp and mould. Unhealthy structures, no longer suitable for people to live in – even if they ever were in the first place compared to their former life in settlements that had been organised around hunting, fishing, kin, and friends, despite the major improvement their designers claimed them to be – their decaying presence has now come to symbolise how Danish modernisation policy was a period of post-colonial urban enclosure and dispossession. Blok P was demolished in 2012, and other lettered and numbered apartment blocks are currently being demolished. As I discuss further in Chapter 3, Nuuk's colonial and post-colonial character is being unmade, while the city is being rebuilt. Many former residents of those old blocks – including Blok P – have been rehoused by the Greenland government and the municipal authorities elsewhere in the centre of Nuuk and in new apartments in the growing suburb of Qinngorput, which is 5 kilometres from the city centre and has been in development since the mid-1990s. A number of Blok P residents expressed their unwillingness to move to Qinngorput, as they would be further from the central part of Nuuk and removed from the immediate vitality of their social networks.

It was the case, though, that some Indigenous Greenlanders accepted and even welcomed the changes implemented initially after colonial status was ended. This was a point many of those in Nuuk with whom we worked and whom we interviewed as part of Inuit Pinngortitarlu wanted to emphasise when we spoke with them about their recollections of movement and resettlement. At the time, while

Figure 2.2 Apartment blocks in Nuuk and other towns on the west coast were built rapidly by the GTO in the 1960s to house people who were moved from small settlements (Photograph Mark Nuttall).

people were being moved from their homes, there was Greenlandic participation – mainly of people drawn from the élite that could trace its origins to the late eighteenth and early nineteenth centuries – in the commissions that scoped out the development policy processes. However, for many others, the nature of the social changes and upheavals experienced following World War II, and specifically in the 1950s and 1960s, also meant a reappraisal of Danish action and was one reason for the emergence of Inuit political parties. There was a heightened sense, awareness, and expression of Inuit ethnic and cultural identity, a desire for self-determination, and a quest for Indigenous sovereignty. A deeper critique of Denmark as a colonial power led to calls for an examination of its post-colonial intent and the implementation of policies of modernisation. For many, decolonisation was not working in a way that was giving Greenlanders more autonomy, as it was supposed to have done now they were Danish citizens. Instead, they were affected by Danish policies on a scale that had never been experienced before, even during the assertive actions of the KGH as a dominant power when it controlled the trade and held sway over the Greenlandic population. With the majority of the Inuit population now living in the fast-growing west coast towns, this demographic transition brought its own considerable social, cultural, and economic problems (Nuttall 1994).

While life in small settlements had been characterised by and organised around family groups and close social association – and attuned to a seasonal round of hunting and fishing – the move to towns and the transition to wage-earning disrupted kinship relations and eroded the patterns of sharing that were characteristic of Inuit hunting and fishing activities. For many people this led to alienation, isolation, social and economic marginality, discrimination, and trauma (Nuttall ibid.). During our work with them, Kaaleeraq and Marius, for example, described this as a difficult time for those who were now living in apartments in Nuuk and who had been separated from surroundings in which they were immediately dependent on maintaining relations with the environment and with animals. They were faced now with negotiating other kinds of relationships with their managers and co-workers in the fish processing factories in which they worked and with new neighbours in the apartment buildings. This was aggravated by ethnic tensions and the presence of the increasing numbers of Danes living in Greenland because of the need for construction workers, teachers, doctors, and administrators (Nuttall ibid.). In some places, such as the coal mining settlement of Qullissat, which had been established in 1924 on Disko Island to produce fuel for the larger settlements along the west coast, people developed a reputation for leftist sympathies, with Greenlandic miners finding inspiration in the global workers' movement, and this spilled over into the political life of the new towns. The Greenland Provincial Council voted to close the mine in 1966 because of falling demand for coal – it was eventually closed in October 1972 and some 500 people who still lived there were relocated to other towns in the Disko Bay region and elsewhere on the west coast. This closure precipitated further political grievance.

The Inuit Party, established in 1963, was the first Greenlandic political party to have wide public support. It was a political reaction to the discrimination that was inherent in Danish policies, such as the controversial 'birthplace criterion' (*fødestedskriteriet*) that meant civil servants born in Greenland received 85% of the wages that were paid to civil servants who were born in Denmark. It was argued by the Danish administration that Greenland would be seen by many who went there as a 'hardship posting' and so financial incentive and reward would be needed to encourage them to move there, especially if they were taking their families with them and if they were to afford to live well. Not only did Danes receive higher wages, they also had access to better quality housing. In effect, the housing programme that had been implemented to improve living conditions by constructing modern dwellings in the growing industrial towns of the west coast was seen as only being of benefit to Danish employees who were moving to Greenland (Petersen 1995).

The birthplace criterion undermined the notion that the end of the colonial period meant that Greenlanders were to be equal citizens with Danes. Instead, they experienced social and economic inferiority to them, often expressed as *ajukkunneq*, and feelings of being enclosed in a hierarchical system that continued to support structures and institutions of colonisation. This was also apparent in the new school system – previously, schools were closely tied to the church (and Greenlandic catechists), but the introduction of a Danish style of teaching by teachers who had attained a high level of education in Denmark and who were not Greenlandic-speakers soon meant that Danish was privileged over Greenlandic as a language of instruction and educational success. Denmark began to send more people to Greenland to work in administration and emerging businesses, and they occupied higher positions in the new economic system. Inspired by Icelandic independence in 1944 and Home Rule in the Faroe Islands in 1948, the Inuit Party advocated for Greenland's complete independence from Denmark, but it failed to maintain momentum on this issue and was dissolved in 1967. In recent years, the controversies, legacies, human rights abuses, and trauma of Denmark's colonial and post-colonial policies implemented in Greenland – especially the resettlement of people and the removal of children from their families (e.g. see Bryld ibid.; Thiesen ibid.[1]) – have been under scrutiny through the work of the Greenland Reconciliation Commission (*Saammaateqatigiinnissamut Isumalioqatigiissitaq*; see www.saammaatta.gl), which was appointed by the government of Greenland (the Danish government did not participate in the commission's work) and released its final report in December 2017.

Under the terms of its formal incorporation into the Danish political state in 1953, Greenland elected two members to the Danish parliament, who from the start represented Indigenous interests. In 1969, Knud Hertling, who had been elected a Greenlandic member of the Danish Parliament in 1964 as a Social Democrat, founded the short-lived Sukaq Party. More enduring and influential was Siumut, which means 'forward.' Organised as a political movement in 1975, it was established as a political party in 1977. The other main parties formed in the 1970s were Inuit Ataqatigiit (IA), which is commonly translated as 'community of the people' and was established formally as a political party in Aasiaat in West Greenland in 1976 – originally inspired by radical socialist ideas, IA took an explicit anti-imperialist and anti-colonial stance – and Atassut, meaning 'link,' 'cohesion,' or 'solidarity,' a liberal-conservative and unionist party in April 1978.

Siumut's origins can be traced to the beginning of the 1970s when Greenlandic politicians Moses Olsen (who had been elected a Greenlandic member of the Danish Parliament), Jonathan Motzfeldt, and Lars Emil Johansen (who were elected to the Greenlandic Council) decided to establish their own political group. Their action was a response to Danish policies enacted in Greenland, especially in the previous two decades, but the beginnings of the Siumut movement can also be traced to a more proximate issue – Greenlandic opposition to membership of the European Community, in part because of concerns over fishing rights, when Denmark joined in 1973 (see below). Siumut has taken something of a moderate socialist position since its establishment and has remained the dominant political party in Greenland, even if its periods in power have been as part of coalition government arrangements. At the election of April 2021, however, this dominance was interrupted when IA came to power and formed a coalition with the centrist-populist Naleraq (IA had previously been in power from 2009 to 2013, when it formed a coalition government with the Demokraatit party). The Inuit Party was, in a sense, the forerunner of IA, which had started as a political movement by Greenlandic students in Denmark in the 1970s. The Inuit Party should not be confused with a breakaway group called Partii Inuit, made up of former IA members, that was in a coalition with Siumut and Atassut in 2013–2017.

Self-government and possible independence

The social and economic transformations initiated since the 1950s, as well as the policies of assimilation and modernisation imposed by Denmark, contributed to the politicisation of Greenlandic Inuit culture, social, and cultural expressions of Indigenous identity (often with reference made to sharing a common pan-Arctic Inuit identity and culture and speaking variations of a common language, with Inuit in Canada, Alaska, and Chukotka), the formation of political parties, and the beginnings of a movement for Home Rule. In the 1970s, Denmark acknowledged there was a growing dissatisfaction felt by Greenlanders as a result of G-50 and G-60 and recognised that a change in its relationship with Greenland within the Danish Realm was necessary. A Home Rule Commission was set up in 1975, followed by the passing of the Home Rule Act three years later. Greenland Home Rule was established by referendum in January 1979 and the election for the first Greenlandic parliament was held in April of that year.

Home Rule recognised Greenland and its people as a distinct community within the Kingdom of Denmark. Legislative and administrative powers in a large number of areas and public institutions were quickly transferred to the Home Rule authorities – the areas Greenland took control of included the economy, taxation, fisheries, hunting and agriculture, the environment, education, the health system, social affairs, and housing. In the matter of trade and the economy, the Home Rule authorities took over control of the KGH and it was reorganised in 1986 as Kalaallit Niuerfiat (KNI – or Greenland Trade, in English). Greenland left the European Economic Community (EEC) in January 1985. It had joined with Denmark in January 1973 following a referendum in 1972, during which a majority of voters in Denmark had been in favour of membership, with a majority of voters in Greenland against it. Greenland's decision to leave was mainly over disagreements with the Common Fisheries Policy, but the government negotiated Overseas Countries and Territories Association (OCTA) status, which allows favourable access to European markets (e.g. a fisheries agreement between Greenland and the European Union is renegotiated every five years).

In 2004, a Danish-Greenlandic Self-Rule Commission was established to negotiate the terms for greater self-government. On 25 November 2008, 75.5% of those who voted in a referendum on self-governance favoured greater autonomy, and on 21 June 2009, the new political arrangement of Self-Rule was instituted. On that morning, I stood with several hundred people making up a large crowd at the old harbour in Nuuk and listened to then-premier Kuupik Kleist, who was heading an IA-led coalition government talk about his hopes for a bright future in which Greenlanders would have greater autonomy and shape their own lives and nation through self-determination. Kleist and other politicians were aware at the time that resource development would likely influence how that future would turn out to be. The Act on Greenland Self-Government is an extension of the powers enacted in the 1979 Home Rule Act and allows Greenland to request that further powers be devolved from Copenhagen to Nuuk in areas that are currently under the control of Denmark, such as the justice system, police system, prison affairs, and the coastguard. Significantly, the Self-Rule agreement goes beyond the Home Rule Act in recognising that Greenlanders are a nation with an inherent right to political independence if they choose it – a further step in the process of decolonisation. The Greenlandic Parliament – *Inatsisartut* – convenes in autumn and spring and has 31 members who are elected for a period of four years; *Naalakkersuisut*, the Greenlandic government, is a ten-member cabinet chaired by the premier.

Greenland is self-governing in domestic affairs, but the following areas of responsibility cannot be transferred to the Greenland authorities until Greenland becomes an independent state: the constitution; foreign affairs (with the modifications and exceptions mentioned in the Legal Act no. 473, Chapter 4); defence policy and national security; the High Court of Justice; citizenship; and monetary and exchange rate policy (Act on Greenland Self-Government, Act no. 473). In 2005, the Danish Parliament had granted Greenland limited statutory powers to negotiate some international agreements on behalf of the Kingdom of Denmark. The Self-Rule Act went further in allowing Greenland to open offices of representation to deal with matters of trade and other areas it has responsibility for.

Greenland has already started to represent itself on the international relations stage, with a diplomatic representative presence in Copenhagen and the EU in Brussels and with offices in the Danish embassies in Washington DC (opened in 2014, it also deals with Canada) and Reykjavik that have opened in the last few years. In November 2021, Greenland opened a representation in the Danish embassy in Beijing. Iceland and the US have consulates general in Nuuk and a number of other countries such as Canada, the UK, Germany, France, Belgium,

Norway, Sweden, Finland, Czechia, and South Korea have honorary consuls, while the European Commission announced in October 2021 that it would be opening an office in the capital. This does not mean direct diplomatic representation, but these missions are concerned with specific relations with Greenland and with their citizens who may be living, working, or travelling in Greenland.

While European and Asian links are important, especially for consolidating markets for fish and shrimp exports and for attracting tourists to Greenland, the Greenlandic authorities consider it natural that they would want to strengthen relationships with the US, Canada, and Iceland. Greenlanders already travel to those countries for education and work opportunities. There is a Nordic connection with Iceland, which has a small Greenlandic community – and a number of Icelanders live and work in Greenland. In October 2020, the US and Greenland signed an agreement on developing closer economic relations as well as co-operating on security matters (see Chapter 4 for further discussion on this), while in October 2022 Greenland and Iceland signed an agreement in Reykjavik to co-operate on issues such as climate change, the development of the green economy, and education and research. As I will discuss in Chapter 4, Greenland is strengthening bilateral relations with the US, but its relationship with Washington DC continues to be framed, influenced, and shaped by American military presence at Thule Air Base (now called Pituffik Space Base) and issues related to geo-security.

In many ways, Greenland is Nordic in character in terms of its social and economic institutions. And, of course, it is part of the Kingdom of Denmark and Greenlanders are Danish citizens. Geographically, however, Greenland is part of the North American continent and shares Indigenous ancestral and contemporary cultural connections with Inuit in Canada and Alaska. These connections are held together tightly through the work of the Inuit Circumpolar Council (ICC), which is a pan-Arctic Indigenous peoples' organisation representing the rights of Inuit in Greenland, Canada, Alaska, and Siberia. Established in Alaska in 1977, in response to increased oil and gas exploration and development in the Arctic and Inuit concerns over the social and environmental impacts, ICC has had non-governmental (NGO) status at the United Nations since 1983. ICC has set about challenging the policies of governments, multinational corporations, and environmental movements, and has argued that the protection of the Arctic environment and its resources should recognise Indigenous rights and be in accordance with Inuit tradition and cultural values. ICC-Greenland's office is in Nuuk.

Kalaallisut is now the country's official language, and debates about making English (rather than Danish) the first foreign language Greenlanders learn in school often centre around discussions that are concerned with how Greenland will meet the challenges of globalisation and the country's place in the wider world. Greenland's constitutional commission (*Tunngaviusumik Inatsisissaq pillugu*) was established in April 2017 and is tasked with preparing a draft of a constitution that would be the legal basis for an independent sovereign Greenlandic state.

Despite the challenges ahead and irrespective of whether it will lead to eventual independence, Self-Rule is a form of governance that allows for the expression of a growing cultural and political Greenlandic confidence that has become more strident in recent years. Greenland has often been looked upon and admired as a model for Indigenous self-government, but it is important to recognise that Home Rule was a process of nation-building that allowed for devolution. Rather than an ethno-political movement, it resulted in a public government arrangement and Self-Rule has moved the nation-building process along a trajectory of self-determination towards state formation (Nuttall 1992, 2017). I would argue that the relevance of Greenlandic self-government goes beyond that of self-determination for Indigenous peoples, however, and says much about the aspirations for autonomy in small political jurisdictions and stateless nations. Scottish nation-builders, for instance, look to Greenland's example as much as Indigenous peoples elsewhere in the Arctic and other parts of the world do.

One of the significant aspects of these processes of nation-building and state formation, however, has been the emphasis on the nurturing of a Greenlandic nation with a shared Greenlandic identity (Nuttall 1992). Bærenholdt (2007) argues that Greenland's political and economic approach since the early 1990s has been to follow a course similar to the Icelandic national-liberalist policies of giving national independence a much greater priority than to the development of village and settlement life, culture, and economy. The economic significance of what is often regarded as a traditional way of life based on hunting and fishing that has sustained small communities is steadily decreasing. There is a noticeable widening divide between town and settlement life. It can be argued that there is a process of internal colonialism in modern, self-governing Greenland. Supportive of greater urbanisation, politicians are often accused of subordinating the villages and settlements as well as the populations of East and North Greenland to the idea of one nation (Bærenholdt ibid.).

Many difficult fiscal issues dominate Greenlandic politics and many questions are unresolved concerning how the country will pay for the responsibilities it is assuming control over and decision-making from the Danish state, as well as how it will lay the foundations for a sustainable economy. Some 60% of Greenland's budget revenue comes in the form of an annual 3.5 billion DKK (around €470 million and US$514 million) block grant it receives from Denmark, as well as Danish state expenses of roughly 1.2 billion DKK (roughly equivalent to €160 million and US$175 million). Greenland's economy remains highly dependent on exports of cold-water shrimp and fish (mainly Greenland halibut, although cod, snow crabs, and mackerel are also important catch species). The main challenge to securing greater self-government and economic independence is to replace the block grant and other Danish state transfers with revenues generated from within Greenland. This requires the development of new economic initiatives, industries, and commercial enterprises as well as investment in education and training. Greenland also recognises that it will continue to need to import skills and expertise, and difficult issues related to immigration are often topics of parliamentary debate as well as for the national media – they often recall the events of the 1960s when Danish workers moved into Greenland in large numbers. In short, aspirations for independence would require Greenland to embrace globalisation further.

When the Danish-Greenlandic Self-Rule Commission was established in 2004, it was also tasked with considering Greenland's claim to mineral rights, its ownership of subsoil resources, and right to the revenues from non-renewable resource development. In 1935, a royal decree had defined mineral resources as the property of the Danish state, and a comprehensive mining policy was introduced in 1963 and a mining act was passed in 1965 (Sinding 1992). The commission concluded that minerals in Greenland's subsoil belong to Greenland and that the country has a right to their extraction – something that Denmark had refused to consider during the discussions about Home Rule in the mid-1970s.

In 1973, the Danish government had granted 13 concessions for oil exploration on the Southwest Greenland continental shelf to a group consisting of one Danish and 19 foreign companies (Taagholt 1978). When negotiations took place for Home Rule in 1977–1978, oil development and mining policy became a primary focus, although a Greenlandic public protest campaign (which also took place in Copenhagen) against offshore exploratory drilling for oil and gas in 1976 and 1977 made politicians in both Greenland and Denmark cautious about expressing their support for hydrocarbons. Extractive industry in Greenland proved to be divisive. Some Greenlandic negotiators wanted full control of mineral resources, while others – mainly Danish negotiators – suggested other arrangements, such as some mineral revenue going to Denmark to recover some of what had been spent on Greenland during the modernisation process of the 1950s and 1960s. In the end, the Home Rule Act did not grant Greenland subsurface rights, although there was a vague description in it of non-renewable resources being viewed in relation to the 'fundamental rights' Greenlanders had to them (Mineral Resources Administration 1989). During the discussion and negotiations that led to the Home Rule agreement, there was a compromise of sorts in relation to an amendment of the 1965 Mining Act, with the sharing of decision-making over resources between the Danish government and the Greenland Home Rule government and the establishment of a joint Danish-Greenlandic parliamentary committee to monitor resources (Sinding ibid.).

In May 1990, the Mineral Resources Administration for Greenland published a report which focused on a strategy for mineral resource development. Mining was considered as an industry with the potential to make a notable contribution to the economy of Greenland (Sinding ibid.). A new Mining Act was passed in 1991. It outlined an overall framework for specific concessions and permits, tenure, taxation, and government-investor relationships. Importantly, it specified that mining permits could only be granted to companies registered in Greenland. In aspiring to develop a mining industry that would create employment and provide tax revenue benefits for the country, Greenlandic politicians pursued an extractive imperative. During the rest of the decade and into the first years of the new millennium, mining policy highlighted the need to increase exploration.

Much of the debate prior to the referendum on self-rule centred on what Greenland's ownership of lucrative resources would mean for greater autonomy and possible independence. With Self-Rule initiated, Greenland took control over subsurface resources on 1 January 2010, and this paved the way for direct negotiation between the Greenlandic authorities and companies interested in exploring and developing resource projects (see Chapter 3). Under the Self-Rule agreement, it was decided that any income generated by subsurface resource development would be administered by Greenland, with the level of the block grant being reduced by Denmark by an amount corresponding to 50% of the earnings from minerals and energy extraction once they exceeded 75 million DKK. The agreement set out how future revenues from oil and mineral resources would then be divided between Greenland and Denmark, while the plan was for the annual block grant to be reduced further and eventually phased out. At the time, it was felt that oil and gas production and mining raw materials would ease this dependence and provide a revenue stream in place of Danish funding.

Naalakkersuisut launched a new oil and mineral strategy in 2014 to cover the period up to 2018. As Bjørst and Rodon (2022) point out, it had the optimistic tagline "Our raw materials have to create prosperity" (Government of Greenland 2014). They argue that it was preparing Greenland for a resource boom that did not in the end happen. Even though venture capital was, at the time, starting to leave Greenland because of market conditions, the strategy chose instead to overlook this fact and present "a much more upbeat vision for the future" (Bjørst and Rodon ibid.). A new strategy had since been introduced to cover the period from 2020 to 2024 (Government of Greenland 2020) and despite the mining and hydrocarbons bonanza that was envisioned in the first strategy not being realised, mining remains an ambition for Greenland's political leaders.

A poll carried out in 2016 by HS Analyse for the Greenlandic national newspaper *Sermitsiaq* suggested there is strong support for independence, although those surveyed who were in the 18–29 age group were more sceptical about whether it could and should be achieved than people in their 60s. However, while the results of an opinion poll conducted by the Universities of Copenhagen and Greenland in 2019 indicated that some 67% of Greenlanders were in favour of independence, 46% of respondents said they would not vote for independence if a referendum were to be held within the next ten years. Aspirations for greater autonomy are seemingly tempered by concerns that it may be another generation or two before independence may be realistic or even desirable. There are many who doubt that mining will bring in the revenue needed for Greenland to forge a sustainable economy that will form the basis for political independence. Fewer people, it seems, would be prepared to accept independence if it meant lowering living standards and ushering in a period of economic hardship.

Note

1 Helene Thiesen's first-hand account of being one of 22 Indigenous Greenlandic Inuit children who were sent to Denmark in 1951 as part of a social 'experiment' in forced assimilation, was translated by Stephen James Minton from the Danish version of her book, *For flid og god opførsel: vidnesbyrd fra et eksperiment* ('For Diligence and Good Behaviour: Testimony from an Experiment', published in 2011 by Milik Publishing; see Minton's translator's forward). Danish social worker and writer Tine Bryld's book, *I den bedste mening* ('With the best of intentions') is a powerful work that was published in 1998. It gave voice to the experiences of those 22 Greenlanders, all of whom were six or seven years old at the time of their removal from their families, and became the basis for the 2010 Danish film, *Experimentet* ('The Experiment'). Directed by Louise Friedberg, *Experimentet* focuses on the opening of an orphanage in Nuuk to turn 16 Greenlandic children into Danes after spending 18 months in Denmark.

References

Bjørst, Lill Rastad and Thierry Rodon. 2022. "Progress stories and the contested making of minerals in Greenland and northern Québec." *The Extractive Industries and Society* 12, 100941. https://doi.org/10.1016/j.exis.2021.100941.

Bryld, Tine. 1998. *I den bedste mening.* Copenhagen: Gyldendal.

Bærenholdt, Jøregen Ole. 2007. *Coping with Distances: Producing Nordic Atlantic societies.* Oxford and New York: Berghahn.

Fraser, Emma. 2018. "Unbecoming Place: Urban imaginaries in transition in Detroit." *Cultural Geographies* 25 (3): 441–458.

Gad, Finn. 1973. *The History of Greenland II: 1700 to 1782.* Montreal: McGill-Queen's University Press.

Gad, Finn. 1982. *The History of Greenland III: 1782–1808.* Montreal: McGill-Queen's University Press.

Gearheard, Shari Fox, Lene Kielsen Holm, Henry Huntington, Joe Mello Leavitt, Andrew R. Mahoney, Margaret Opie, Toku Oshima and Joelle Sanguya (eds.) 2013. *The Meaning of Ice: People and sea ice in three Arctic communities.* Montreal and Hanover: International Polar Institute.

Government of Greenland. 2014. *Our Mineral Resources – Creating Prosperity for Greenland: Greenland's Oil and Mineral Strategy 2014–2018.* Nuuk: Ministry of Labour, Industry and Trade.

Government of Greenland. 2020. *Greenland's Mineral Strategy 2020–2024.* Nuuk: The Ministry of Mineral Resources. February 2020, https://govmin.gl/wp-content/uploads/2020/03/Greenlands_Mineral_Strategy_2020-2024.pdf.

Grydehøj, Adam. 2014. "Constructing a centre on the periphery: Urbanization and urban design in the island city of Nuuk, Greenland." *Island Studies Journal* 9 (2): 205–222.

Gulløv, Hans Christian. 1997. *From Middle Ages to Colonial Times: Archaeological and ethnohistorical studies of the Thule Culture in South West Greenland 1300–1800.* Meddelelser on Grøland, Man & Society 23. Copenhagen: The Commision for Scientific Research in Greenland.

Hamilton, Lawrence, Per Lyster and Oddmund Otterstad. 2000. "Social change, ecology and climate in 20th century Greenland." *Climatic Change* 47 (1–2): 193–211.

Hayashi, Naotaka. 2013. *Cultivating Place, Livelihood, and the Future: An ethnography of dwelling and climate in Western Greenland.* PhD thesis, Department of Anthropology, University of Alberta.

Heinrich, Jens. 2018. "Independence through international affairs: How foreign relations shaped Greenlandic identity before 1979." In Kristian Søby Kristensen and Jon Rahbek-Clemmensen (eds.) *Greenland and the International Relations of a Changing Arctic: Postcolonial paradiplomacy between high and low politics.* London and New York: Routledge, pp. 28–37.

Hermann, Anne Kristine. 2021. *Imperiets Børn.* Copenhagen: Lindhard og Ringhof.

Kaasa, Adam. 2020. "Domestic monumentality: Scales of relationship in the modern city." In Jonathan Bach and Michał Murawski (eds.) *Re-Centering the City: Global mutations of socialist modernity.* London: UCL Press, pp. 245–252.

Lennert, Ann Eileen. 2017. *A Millennium of Changing Environments in the Godthåbsfjord, West Greenland: Bridging cultures of knowledge.* PhD dissertation. Nuuk: Ilisimatusarfik/University of Greenland.

Mineral Resources Administration. 1989. *Report from the Strategy Group for Mineral Resources in Greenland.* Nuuk: Mineral Resources Administration.

Nuttall, Mark. 1992. *Arctic Homeland: Kinship, community and development in Northwest Greenland*. Toronto: University of Toronto Press.

Nuttall, Mark. 1994. Greenland: Emergence of an Inuit homeland. In Minority Rights Group (ed.) *Polar Peoples: Self-determination and development*. London: Minority Rights Group, pp. 1–28.

Nuttall, Mark. 2012. "The Isukasia iron ore mine controversy: Extractive industries and public consultation in Greenland." *Nordia Geographical Publications* 41 (5): 23–34.

Nuttall, Mark. 2017. *Climate, Society and Subsurface Politics in Greenland: Under the Great Ice*. London and New York: Routledge.

Petersen, Robert. 1995. "Colonialism as seen from a former colonized area." *Arctic Anthropology* 32 (2): 118–126.

Priebe, Janina. 2015. "The Arctic scramble revisited: The Greenland Consortium and the imagined future of fisheries in 1905." *Journal of Northern Studies* 9 (1): 13–32.

Priebe, Janina. 2017. *Greenland's Future: Narratives of natural resource development in the 1900s until the 1960s*. Umeå: Umeå Universitet.

Rose, C.H., P. Nasen, R. Jess Jørgensen and D.E. Jacobs. 1984. "Sheep farming in Greenland: Its history, managemental practices and disease problems." *Nordisk Veterinærmedicin* 36 (3–4): 65–76.

Simmel, Georg. 1950. *The Sociology of Georg Simmel* (Translated and edited by Kurt H. Wolff). Glencoe, IL: The Free Press.

Sinding, Knud. 1992. "At the crossroads: Mining policy in Greenland." *Arctic* 45 (3): 226–232.

Sørensen, Axel Kjær. 2007. *Denmark-Greenland in the Twentieth Century*. Meddelelser om Grønland, Man & Society 34. Copenhagen: Museum Tusculanum Press.

Taagholt, Jørgen. 1978. "Arctic resources." *Arctic Bulletin* 2 (14): 347–352.

Thiesen, Helene. 2023. *Greenland's Stolen Indigenous Children: A personal testimony* (translated by Stephen James Minton). London and New York: Routledge.

3   Re-making and becoming

When reserves of oil, gas, and minerals become depleted – or if political circumstances make existing operations of production in sensitive areas more volatile and exposed to instability – extractive industries typically plan moves to expand their reach into other places of potential that are often viewed by states and companies as new terrains and waterscapes of resource extraction. In this process, imaginaries of resource frontiers and hopes for the discovery of resources are persistent. They are also seductive in how spaces of extraction are marked out and represented in business-speak as promising significant yield and value for investors in resource projects (Cons and Eilenberg 2019; Luning 2018).

Increasingly, this process includes expansion into areas of the planet that are viewed by industry as 'marginal,' 'remote,' and 'peripheral' spaces, such as in the Arctic. Historically, they have been represented and given form as empty places and imagined as 'last frontiers' that are rich in 'undiscovered' resources, even though they are often homelands for Indigenous peoples and local communities (Nuttall 2010; Stammler and Ivanova 2016). Extractive resource companies are also exploring the prospects for hydraulic fracturing in what were previously regarded as geological formations that would only likely contain economically unviable or unconventional oil and gas resources thought of as difficult to extract (De Rijke 2017; Willow and Wylie 2014), or are investigating the possibilities for seabed mining (Childs 2020; Zalik 2018). "As key resources become scarce," write Richardson and Weszkalnys (2014: 5):

> new resources come into existence. Across the globe, states and corporations have redoubled efforts to extract conventional and unconventional resources in an attempt to deliver ongoing prosperity to citizens and shareholders. The contradictions and violence of these endeavors are most apparent in state-sanctioned encroachment of multinational companies on indigenous and other rural lands.

The sharpened interest expressed by mining and oil companies in Greenland over the last couple of decades has given rise to ideas and political and economic narratives related to what Gavin Bridge (2004) calls bonanza geographies. By this, Bridge refers to places in which investment from the extractive industries sector and activities of exploration and development appear to be especially influential

DOI: 10.4324/9781003175421-4

and powerful in how they are put to work to transform them into lucrative re-source spaces. In Greenland, these terrains and waterscapes of extraction have become part of contemporary political life. The subsurface (including the seabed), and what it promises to yield, figures in economic calculations and political deci-sion-making for a prosperous future, just as it has for many other countries expe-riencing, or on the verge of, resource booms. Greenland's mineral strategies have continued to emphasise that raw materials have to create prosperity (Government of Greenland 2014, 2020). The current mineral strategy states that:

> The overall objective of the Government of Greenland for the mineral re-sources sector is to develop a leading industry which will contribute positively to the economic development and create new jobs for the benefit of all of Greenland. In other words: Greenland is to be an attractive mining country which investors will prefer over other mining countries.
>
> (Government of Greenland 2020: 8)

Resource economist Erich Zimmermann (1951) articulated the basis of an ar-gument that has guided much scholarly reflection on the nature of resources and their extraction. Resources, he said, are not already there in the ground ready to be unearthed, they 'become,' as do the places and spaces from which they emerge. Resources and the territories in which they are located are not fixed, nor are they finite. Becoming, then, is as much an ideational process as it is one grounded in material or physical aspects (De Gregori 1987). It is not enough, however, for mineral resources to be discovered and assessed in order to attract the attention of global investors. Resources need to be converted into profitable *mineral reserves* which "imply a higher level of knowledge of the resources and therefore, a high level of confidence for the potential investor" (Ayeh 2021: 156).

Only a few years ago, the near future (which has now arrived as the present) was imagined to be one in which a number of mining projects were up and run-ning in Greenland; indeed, this was the objective of Greenland's mineral strat-egy for 2014–2018. Although I had been interested in extractive industries in Greenland for some time and had been following discussion over oil exploration, mining, and megaprojects such as an aluminium smelter and its associated in-frastructure for hydropower production that was being planned by Alcoa near Maniitsoq, as well as the early work by Greenland Minerals on a planned uranium mine at Kuannersuit (Kvanefjeld) near Narsaq, I decided to embark on specific research in Nuuk in February 2012. This, I thought, would focus primarily on the politics, bureaucracies, narratives, and imaginaries surrounding mining and oil exploration, and how all this contributes to the making of resource spaces. I was interested in the excitement about what was about to happen. It was also an interesting time, because even then it was beginning to look as if a number of pro-jects that were being promoted by companies and being considered for approval by the Greenlandic authorities concerned with the administration of the resource industry would not actually happen – despite appearances to the contrary and the expressions of hope I heard.

Resource talk

I began to speak with politicians and business leaders about the prospects for the development of an extractive sector. Indeed, extractive industry was a topic that was difficult to avoid. It seemed that everyone was eager to talk about how optimistic they were that Greenland was on the verge of a minerals and oil bonanza, even though most admitted that it would bring significant social, environmental, and regulatory challenges (Nuttall 2017). "Greenland is an unexplored frontier, full of wealth in the ground" is how one Danish construction company owner from Nuuk told me in spring 2012. At the time, he had lived in Greenland for over twenty years and was married to a Greenlandic woman. He owned a boat and fished in Nuup Kangerlua whenever he could. "There's just so much empty space here," he said. "There's lots of room for mines to be developed." He was hopeful about resource development and had started to make plans for how his company could provide services for international companies, mainly with equipment and labour. One afternoon, we sat in his kitchen drinking coffee and he gestured to the view from the window: "Life is good, here in Nuuk," he said. "Oil and mining will make it even better. That's the way for Greenland to make money. But, it has to be done the right way." Former Greenlandic premier Kuupik Kleist told me in early 2010 that resource extraction presented quite a considerable dilemma to his government: mining and at that point oil development seemed unavoidable and desirable, he said, given the nature of the Self-Rule Act and Greenland's pursuit of an extractive imperative and the increasing confidence expressed that the country could become independent within a few years – yet, he reflected that some hard decisions would need to be made and not all would necessarily be in the best interest of society and the environment. Kleist was an IA politician, presiding over a coalition that was in power when Self-Rule was introduced in June 2009, and I had given a talk at a meeting in Nuuk that he and a few members of his government attended on how the regulatory process for extractive industry worked in Canada. Greenland was then, and still is, working out the best direction to take regarding regulation and administration. At the time of my initial conversations with people in Nuuk about resource extraction, mining companies were already busy prospecting in several parts of the country, oil exploration was taking place in west coast waters, and seismic surveys were underway in northern Baffin Bay. Companies held information sessions to give the public an opportunity to hear about their plans, but the format often fell way short of in-depth discussion and consultation. Each week brought a new delegation from an oil or mining company to Nuuk to hold meetings with the Bureau of Minerals and Petroleum (BMP). They also wanted to talk to scientists, archaeologists, and economists. Given my affiliation to Ilisimatusarfik/University of Greenland and the Greenland Climate Research Centre, I was often invited to sit in on discussions about their plans. I was also involved in meetings held by Indigenous organisations such as ICC-Greenland, occupational groups such as the Association of Fishers and Hunters in Greenland (KNAPK), and local environmental NGOs, about the environmental and social impacts of extractive industry. I was also interviewed by Greenlandic television, radio, and print media about Canada's experience with mining and oil extraction, especially as there was interest in the regulatory process in other parts of the Arctic. And, as I discussed in Chapter 2, the 'Inuit Pinngortitarlu' project was getting started and a sharp focus of that work was understanding the extent of mineral prospecting and how exploration and exploitation licences were overlapping with customary use of Nuup Kangerlua, and hearing about the concerns people had that mining would affect reindeer hunting, seal hunting, and fishing.

There was significant opposition to mining expressed by local organisations and public interest groups – notably directed at the London Mining iron ore mine project at Isukasia and the uranium extraction venture at Kuannersuit in South Greenland (more on which below) – but I was struck by the contrast presented by the excitement and the positive mood in the offices and meeting rooms of government departments, geologists, industry consultants, and entrepreneurs that Greenland was regarded by international interests as an emerging resource frontier (Nuttall ibid.). Caught up in this resource talk, Greenland was being visualised as a dynamic space within which mountains, remote fjords, oceanic trenches, ridges, fractures, and fissures were waiting to be explored and their potential as mineral and hydrocarbon reserves assessed (Dodds and Nuttall 2018). Much of this talk was concerned with speculation as an observation of potentiality, both in the way that the very existence of potential is remarked upon and in the way that facts are constantly produced about it (see Weszkalnys 2015). People in government and industry with whom I spoke emphasised the diversity and richness of Greenland's geology and the chances of good discoveries (as the current mineral strategy continues to do). I began to look a little deeper into how Nuuk was being reimagined and redesigned as a hub for scientific, technical, and consultative expertise and as a logistics support base for extractive industry. As I was leading a research programme on climate and society in Nuuk with Greenlandic PhD students working on the impacts of climate change as well as the emerging extractive industry sector, I was also well placed – indeed, immersed – in this environment of speculation and planning for the future.

In the years since, an entire industry has grown in the capital and other towns as well as in Denmark, which involves navigating and negotiating a broad range of administrative, technical, and scientific procedures and bureaucracies. Each have distinctive forms of expertise and claims to perform calculative labour devoted to smoothing the way for resource extraction. Creating spaces of exploration and building the infrastructure for mines to unearth resources brings together many different interests and communities of expertise in different places and at numerous sites, not only in Greenland but globally. These range from international companies, investors, politicians, engineers, geologists, environmental consultants, and local enterprises. Extractive industry, then, should not be understood only as the activities of a company that is involved in scoping out terrain, exploring for minerals and hydrocarbons, developing a project, and engaging in the productive tasks of unearthing. It also encompasses the scientific work and the geological assessments behind it as well as the regulatory framework, government bills and acts, regulation proposals, amendments, environmental assessments, and the monitoring of prospecting, exploration, and production. In this way, when

I use the term 'extractive industry,' I refer to the greater context and the political and social dimensions of how projects are imagined, thought about, scoped out, and implemented, and how decisions are made that put the structures and institutions in place that administer, govern, and regulate resource projects, but which also administer, govern, and regulate the political and public environments in which projects emerge, take shape, and are legitimised.

As Mezzadra and Neilson (2017) argue, the intensification and expansion of extractive industries in contemporary capitalism demands an attentiveness not just to the literal productive labour and the earth moving and digging that extraction involves, but to the new forms of extraction that are happening on different kinds of frontiers such as data mining and biocapitalism. Getting to grips with this involves understanding what they call operations of capital as a way of tracing a complexity of dense connections between the far-reaching, expansive logic of extraction and capitalist activity in the spheres of logistics and finance. 'Extractive industry,' therefore, and to reiterate the point above, refers not just to the productive activities companies embark on when they go about prospecting, digging, and extracting minerals from the subsurface or when they develop oil and gas projects, but to a broad range of procedures, bureaucracies, enterprises, forms of labour, and marketing that involve administrators, officials, members of parliament, government ministers, advisors, and consultants. This constitutes and generates an enormous administrative and bureaucratic machinery which engages, puts into action, and oversees different kinds of labour. There is even an imaginative labour concerned with speculation about future projects and their prospects and with economic and environmental assessments. Much of this kind of work that takes place in Nuuk does so in the various agencies and departments that make up the Mineral Resources Authority (for more on this, see Chapter 5) as well as in private consultancy firms, and engages people from various departments of Ilisimatusarfik/University of Greenland, the Greenland Institute of Natural Resources (GINR), and the Greenland National Museum.

Sabine Luning (2018) offers some cautionary and critical remarks about the resurgence of the frontier concept as an analytical tool in studies of resource extraction – it can be used too loosely and is often not well defined, she says – but nonetheless she argues for understanding its role in the politics of representation. Frontier speak complements resource talk. Indeed, the global mining industry talks about Greenland as a new frontier for exploration and production. The object, area, and importance of the resource frontier may vary, depending on the region, but frontier speak sets in motion various processes that create the very thing that it names, along with ideas about bountiful opportunities, challenges, vulnerabilities, and risks (Dodds and Nuttall 2016: 118). As Simon Schama (1995: 7) writes, "wilderness, after all, does not locate itself, does not name itself."

Making and marketing the resource frontier

In my fieldwork in Nuuk, I have been particularly interested in the way that this industry and resource talk makes the emergence of an extractive frontier and the formation of resource spaces in Greenland possible. All of this not only involves economic development planning but the production of geological inventories of the Greenlandic subsurface and environmental reports and impact assessments. Resource projects also involve archaeologists, marine biologists, and ecologists – scientific experts who are tasked with preparing reports and assessments of the state of cultural and historic sites, bird colonies, marine mammal populations, or polar bear habitat. This work is carried out for the production of strategic environmental assessments, oil spill sensitivity atlases and the marking out of special areas of habitat and conservation. I have experience of this through my work at the Greenland Climate Research Centre, travelling with marine biologists on a research vessel in the stormy waters of Baffin Bay to research the possible impacts of seismic surveys on marine mammals, and gathering data that fed into a strategic impact assessment. Places in which oil exploration and mining occur become sites of regulation. Biologists from GINR, for instance, are tasked with doing population counts of marine mammals, fish, birds, reindeer, and musk oxen, and assessments of habitat, while archaeologists from Greenland's national museum are sent into the field to do surveys of Paleo-Inuit and Thule sites and determine whether the areas designated for potential mines are free of the evidence of past human settlement and cultural heritage. Often, this can seem to be an added burden, given their usual daily workload. One archaeologist I know at the national museum once expressed his frustrations to me about the pressure he and some of his colleagues were under to assess sites in areas where mining concessions had been granted to companies by the government. He explained that there were not enough resources to do as much archaeological and cultural heritage fieldwork as they needed to do in the first place, and so being diverted to do such assessment work was taking them away from other projects. He told me that, of course, it is an urgent task to determine how much historic human presence there is at potential mining sites, but that the number of assessments that needed doing far outstripped the capacity he and his colleagues had to do them.

Not all the survey work leads to a recognition that the preservation of cultural heritage is vital, as decisions taken around Alcoa's planned aluminium smelter near Maniitsoq reveal. The Government of Greenland had been in discussion with Alcoa since 2006 to construct this megaproject; by 2016, no agreement had been reached and the plan floundered, making it another example of an unfinished project (Carse and Kneas 2019). The national museum carried out studies that were done as part of the assessment work for the two hydropower plants that would be needed for the smelter. In areas inland called Tasersiaq and Tasartuup Tasia, new knowledge about Paleo-Inuit and Thule culture emerged from the surveys, and the museum recommended to the government that the areas should be protected as sites of cultural historic interest. Ove Karl Berthelsen, who was minister of industry and mineral resources at the time, had this to say:

> A preservation will hinder the use of Tasersiaq as a hydropower resource and thereby remove the foundation for the project. Such an issue may also be relevant with other lakes that can be utilized for hydropower. The cabinet

believes the weighing of preservation and the possible alternate use of areas must be political decisions in the self-rule government's political organs that must determine how the country's resources and development shall be mutually prioritized.[1]

Although the smelter project did not happen, the sites for the hydropower plants are still identified as crucial for Greenland's electricity needs and discussion concerning their construction is ongoing.

At GINR, scientists are required by the Government of Greenland to provide the scientific basis and monitoring work required for assessments of the sustainable use of living resources in and around Greenland for quotas for fishing and hunting as well as for decisions on environmental protection and securing biological diversity. A dedicated Department of Environment and Mineral Resources was established a few years ago to advise Naalakkersuisut on environmental issues related to mineral resources. This includes consultancy related to exploration and exploitation of minerals and offshore activities related to oil and gas exploration. In particular, the department advises the Environmental Agency for Mineral Resource Activities (EAMRA) on matters of concern relating to possible environmental impacts associated with mining activities. The advice is related to the issuing of exploration and exploitation permit licenses, approval of project fieldwork, impact assessments, and monitoring in connection to construction, operation, and decommissioning of mines.

Consultants, administrators, government officials, engineering companies, and financiers as well as documents that include geological inventories of the subsurface, economic development plans, and other techniques of a lucrative supply, services, logistics, and impact assessment sector constitute an administrative apparatus sustained by speculation and estimation about future projects and their prospects. A focus on the construction and operational aspects of extractive projects ignores this broader context, in which subterranean environments are made visible – indeed, how they are thought of and represented as spectacular (cf. Igoe 2017) – and their dimensions become subject to economic calculation and valuation. Nuuk is a rapidly growing capital of almost 19,000 people. As a base for these activities and other forms of innovation from telecommunications and digital infrastructure to scientific research as well as for the distribution of essential goods and supplies around Greenland, Nuuk is not just a centre of calculation, to use Bruno Latour's (1987) notion. It has also become a centre for abstraction (cf. Mason 2022). In this way, Nuuk is very much as part of the planetary mine as a complex, convoluted, and often contradictory geography of extraction (Arboleda 2020).

Resource development is promoted enthusiastically by the Greenland self-rule government, both in Nuuk and at international venues such as major industry conferences and at events announcing bids in new licensing rounds. Greenlandic politicians and representatives of government agencies responsible for regulating extractive projects globetrot each year to Europe, North America, China, and Australia, marketing the country as attractive for (and welcoming of) mining

(and until recently, oil) companies. The mineral strategy addresses the importance of this:

> Greenland competes with other countries to attract exploration investments.
> This makes it an absolute necessity that we contribute to minimising the in-
> vestment risk of the mining companies and thus continue our efforts towards
> becoming an attractive investment destination…. we will also increase local
> and international awareness of Greenland through cooperation. Finally, we
> want to increasingly communicate news about the mineral resources sector
> to the public in Greenland and in the rest of the world.
> (Government of Greenland 2020: 11)

Consummate sales craft and marketing strategies depend not only on the idea
of Arctic minerals as alluring, but of Greenland as an appealing and viable place
in which to prospect, explore, and extract. An obvious irony is that industries
widely linked to climate change – a concern for the Greenlandic authorities despite their desire to see a mining industry developed – are being encouraged to
work in a country that is being seriously affected by global warming. Opportunities for extraction arise – it is often reported – from changing conditions like ice
melt, a process Leigh Johnson (2010) calls "accumulation by degradation."

In August 2013, Richard Mills wrote an article for Mining.com, an online news
site that covers the global mining and minerals industry, about the advantages
offered in terms of a relatively mild climate and open water access to potential
resource spaces in West and South Greenland. He pointed out that:

> In Greenland, if you are working in the southwest part of the country, you
> are never far from year round open deep sea water, meaning a sea port, or the
> possible future site of a rough loading dock, tie up for a couple of barges and
> deep water anchorage for a freighter.[2]

This is something that the mineral strategy emphasises too:

> Many mineral-rich countries have a more well-developed mineral resources
> industry and infrastructure than Greenland. Exploration and extraction in
> Greenland may therefore be associated with relatively high costs compared
> to other countries. However, Greenland's many deep fiords allow for sea
> transport almost directly to and from many of the mineral resource deposits.
> (Government of Greenland ibid.: 9).

But even as I encountered more industry players attracted by this idea and
logistical ease and spoke with them about their interests in Greenland, I had noticed that some mining and oil companies were moving out of the country. Some,
like London Mining and True North Gems went bankrupt; global commodity
prices began to make Greenland less attractive; and renewable energy projects
started to become Greenlandic government priorities. It seems that, just like the

Tokyo host clubs studied by Takeyama (2016), for some companies, Greenland's resource spaces are sites of hope, dreams, aspiration, and longing. They are places of excitement from which something good and worthwhile can be expected from the time spent and invested in looking and trying to wonder what is beneath the promising and seductive layers of the surface, but they are also spaces of desperation. This has prompted actors in resource-focused agencies to work even harder in cajoling them to stay and encouraging others to invest. Success depends on an ability to appeal and entice.

Attraction is key to Greenland's strategising for developing a mining industry, and Bjørst (2020), in considering the emotional labour and hard work that goes into relationship building for mining projects – even to attract interested parties in the first place – argues this involves flirtation to succeed. Mining companies also need to do their bit in enticing investors and winning over communities in support of their projects. Flirtation, however, involves the negotiation of ambiguity (Tavory 2009). Phillips (1994: xii) writes how flirtation keeps things in play and offers the promise of something surprising too. It makes "a game of uncertainty, of the need to be convinced" and "confirms the connection between excitement and possibility, and how we make uncertainty possible by making it exciting." While success depends on the ability to appeal, it also depends a great deal on seduction, in the way Baudrillard (1990: 7) sees seduction as being put into operation through strategies of "play, challenges, duels, the strategy of appearances." It is not a one-way process however, but a mutual relational one.

The anticipated and very much hoped for Greenlandic resource bonanza has not yet happened for many reasons, but mainly because of market conditions, a lack of confidence from investors, and technical and environmental challenges in remote locations. Again, it is worth quoting from the 2020–2024 mineral strategy:

> The former Oil and Mineral Strategy 2014–2018 which (sic) included specific goals for a specific number of producing mines and the establishment of a specific number of offshore oil and gas projects in the strategy period. Unfortunately, we had to realise that things have not gone as expected, among other things because of unfavourable price movements within the mineral resources sector.
>
> (Government of Greenland ibid.: 8)

Some mining projects that looked extremely promising just a few years ago, such as the Isukasia iron ore mine in Nuup Kangerlua, did not in the end get started, even though a thirty-year exploitation licence had been granted to UK-based London Mining in October 2013 to develop and operate it. The project was halted when London Mining went into administration in October 2014. The company no longer had the money to continue with its only active venture, the Marampa mine in Sierra Leone. A slump in iron ore prices had made things difficult for London Mining, but the Ebola crisis in West Africa weakened the company further. In 2015, the project, along with the exploitation licence, was taken over by Hong Kong-based company General Nice, a coal and iron ore importer

(Nuttall 2017). In 2020, the Covid-19 pandemic also affected logistics for mining operations, limiting activities in Greenland, but the government responded by waiving mineral licence obligations and expenditures until the end of that year as part of a relief package to help companies retain their projects.

Critical minerals and sustainable mining

As I will discuss further in this chapter, Greenland's new coalition government that was formed after the election of April 2021 suspended new oil prospecting and exploration licences three months later in July. However, this was probably not a major economic and financial sacrifice to make. As geologist Flemming Christiansen (2021) points out, while there had been significant investment in exploration for hydrocarbons and great expectations for high resource potential for a decade or more from the beginning of the 2000s, most major oil players had already decided they would relinquish their offshore licenses by around 2015, leaving smaller companies to pursue exploratory activities. Despite an initial strong interest in Greenland, Christiansen attributes this to a range of factors, such as a combination of the worries and uncertainties that followed the Deepwater Horizon blow-out in the Gulf of Mexico in April 2010, rapidly increasing shale oil production in the US, a fall in global oil prices, and the introduction of stricter environmental regulations by the EAMRA (see more on this agency below), as well as the difficulties and environmental challenge of operating in ice-filled waters and glacial environments. This move away from Greenland was accompanied by a loss of expertise.

Christiansen, was a former deputy director of the Geological Survey of Denmark and Greenland (GEUS) in Copenhagen. With decades of experience of geological research and now running his own geosciences consultancy, he is widely respected as someone with deep knowledge of oil and gas exploration in Greenland. He is not optimistic that the major companies will return to make long-term investments in offshore Greenland, which they consider to be a high-cost, sensitive environment:

> Attracting industry to invest in exploration in a remote Arctic country like Greenland relies on strong drivers within large companies, in the service industry, and not least within the local political-administrative system. Those hundreds of geologists, geophysicists, petroleum engineers and administrators who worked with Greenland exploration 5–10–15 years ago have all, with very few exceptions, changed jobs to look at other areas or started completely new careers, often dealing with other energy resources. This dramatic loss of corporate memory makes a revival of exploration in Greenland difficult. There is a similar loss of knowledge with the authorities in Greenland.
> (Christiansen ibid.: 1–2)

Christiansen does suggest, though, that there is a possibility that a future Greenlandic government could reverse the oil suspension decision. He feels that some national oil companies would then bid on some offshore blocks for geo-strategic reasons. He thinks however that many Greenlandic politicians prefer onshore exploration rather than offshore exploration – partly because of a conflict with the fishing industry over seismic activities, drilling, and production – and considers it likely that the Nuussuaq peninsula north of Disko Bay and Jameson Land on the east coast will be re-opened for exploration licences.

At times, though, it does seem as if Greenland's resource sector is being built on speculation and hope rather than actual productive activities. Giving rise to an economy of expectation (cf. Weszkalnys 2011), it promises more than it currently yields. Currently, only two mines are operating in Greenland – a ruby and pink sapphire mine near Qeqertarsuatsiaat in Southwest Greenland, which is operated by Greenland Ruby, a Nuuk-based subsidiary that is part of the Norwegian LNS Group (it took over the project from True North Gems), and the Qaqortorsuaq (White Mountain) anorthosite (calcium feldspar) mine southeast of Sisimiut on the central west coast, which is operated by Hudson Resources, a Canadian company. Other projects, although proliferating, are making slower progress, even if some of them are likely to begin eventually. The website of the Mineral Resources Authority has detailed, regularly updated maps that show assigned exploration and exploitation licences (see https://govmin.gl/).

Economics, markets, and logistical challenges determine how mining companies view their prospects. Many consider it only worthwhile to explore and develop ventures in Greenland – which are often in locations that are costly to reach and operate in – if there is a unique mineral deposit. And, of course, having less ice in a fjord does help when shipping ore from an Arctic mine. A dip in global commodity prices and markets over the last few years as well as the effects of other global processes on plans for resource development in the Arctic have made some companies wary about working in Greenland. While this has challenged politicians to think carefully about relying too heavily on extractive industry projects, the view from Nuuk is that mining should still be one pillar on which Greenland builds its economy. In the last year or two, this position has emphasised the development of projects that are focused on critical minerals that can help shape a green energy agenda.

Many minerals that are defined as rare and critical come from a small number of producers operating in only a few countries. For example, in the cases of lithium, cobalt, and rare earth elements, the world's top three producers – China, the US, and Myanmar – control more than three-quarters of global output. China, however, is by far the major producer and exporter of rare earths and is also dominant in the production of other critical minerals.[3] Ensuring access to those critical raw materials necessary for high-technology applications, including renewable energy development and meeting the challenge of climate change but also for the manufacture of consumer electronics, has become an urgent environmental and economic matter. It is also a geopolitical and strategic issue.[4]

As Julie Klinger discusses, in response to China's near dominance of the global production of rare earths, campaigns have been launched "to mine rare earths in the most forbidding of frontiers: in ecologically sensitive indigenous lands in the Amazon, in war-torn Afghanistan, in protected areas of Greenland, in the depths of the world's oceans, and even on the Moon" (Klinger 2017: 3). In more recent years, then, attention has turned to thinking about Greenland as a rare earths frontier. Just as mineral resources such as rubies, pink sapphires, uranium, and rare earth elements have become objects of value for envisioning the shaping of Greenland, they are also influential in shaping how many other nations view Greenland's geostrategic importance. The potential for mining rare earth elements and critical minerals places Greenland at the centre of global discussion on energy security.

The US, China, and a number of other countries too are reported to be positioning themselves to take advantage of Greenland's resource potential, rare earth elements, markets, and geostrategic location. In June 2019, for instance, the US Department of State and the Greenland Ministry of Mineral Resources and Labour formalised a joint mineral sector technical engagement in Greenland through a memorandum of understanding that provided a framework for Greenlandic and US co-operation on mineral sector governance. One initial project was an aerial hyperspectral survey in the southwest of the country that summer. Hyperspectral surveys are a process by which information about minerals and rocks is gathered by measuring sunlight that is reflected from the surface of the ground. It can inform geologists of unrecognised mineralisation and improve geological maps that are of use for mineral exploration.

In December 2020, Iceland's Ministry of Foreign Affairs released a report called *Greenland and Iceland in the New Arctic* in which business and investment opportunities, as well as other areas of co-operation, were identified as emerging from these changing environmental and geopolitical circumstances. The report, which was based on recommendations by a Greenland Committee appointed by the ministry, was widely publicised and garnered significant interest. In particular, it points to the opportunities for developing a mining industry:

> Greenland is in a key position due to the unusually large deposits of usable metals, oil, gas and precious stones in the ground. A major supply of rare earth metals have put Greenland in a new and unique position. Several overlapping factors indicate that the 21st century will be prosperous for the Greenlandic mining industry. Since the financial crisis of 2008, the global inventory of minerals is being depleted, few new mines have been commissioned and a growing shortage of valuable metals is noticeable in the markets. An upswing is therefore anticipated in global mining in the coming decade, including in Greenland.
>
> (Government of Iceland 2020: 140)

On 6 April 2021, mining companies watched nervously, however, as the result of a snap national election saw the left-wing, pro-independence IA party emerge with the most seats in the 31-seat Inatsisartut. IA's 12 seats, however, were not enough to give it a majority – a not altogether unsurprising outcome, as all Greenland's governments since Home Rule was introduced in 1979 have been coalition arrangements. An agreement was made with the centrist-populist, and similarly pro-independence Naleraq party (which won four seats) to form yet another coalition government, with IA's Múte B. Egede as premier (at the time of writing this book, the government had changed to a coalition between IA and Siumut, which was formed in April 2022).

In the run-up to voting day, IA had expressed its continued opposition to the planned project to mine uranium and rare earth elements at Kuannersuit, a mountain near Narsaq in South Greenland. This is, to say the least, a controversial venture that has dominated Greenlandic politics over the last decade. It has often divided public opinion. In October 2013, Inatsisartut had voted in favour of repealing what had been, since 1987, a zero tolerance policy on the mining of uranium and other radioactive elements. It was a narrow win for Siumut, with 15 votes to 14 (two votes were not cast). IA was opposed to ending the ban on uranium mining, but during the zero-tolerance debate, then-premier Aleqa Hammond had stressed the urgency of lifting it for the development of Greenland's economy. She cited low living standards and a high unemployment rate as pressing domestic challenges that could only be dealt with by bringing in revenue from mining – again an expression of an extractive imperative that was guiding Greenlandic political decision-making. With the repeal of the zero tolerance policy, Greenland's government expressed hope that it would pave the way for applications for exploration and exploitation licences for uranium or for mining of rare earth metals, with uranium extraction as a by-product (Nuttall 2013). In this context, it is important to note that Siumut politician Hans Enoksen, who served as Greenland's third premier from 2002 to 2009, had opposed his party's support of uranium mining and left in January 2014 to form Naleraq (which was originally called Partii Naleraq).

Internal political disagreement over the Kuannersuit project and which path to take that could lead to independence led to the spring 2021 election being called. The previous government, a coalition between Siumut, Demokraatit (the Democrats) and Nunatta Qitornai that had been in place since May 2020, became untenable when Demokraatit pulled out of the arrangement in February 2021 (the party has always taken a sceptical view about the prospects of independence). Greenland's then-premier Kim Kielsen from Siumut, which was the lead party in the coalition, had previously been ousted as party leader in November 2020 (although he continued as premier). One of the new government's first moves in April 2021 was to announce that it would ban the mining of uranium and other radioactive-minerals, something IA had promised to do since 2013. This put the Kuannersuit project – and the future of Greenland Minerals, the Australian company hoping to develop it – in doubt. In January 2021, the previous coalition government had awarded two uranium exploration licences to Orano, a French nuclear fuel cycle company, for work in Southwest and Southeast Greenland. In May that year, Orano announced it would be suspending its exploratory activities, but would retain the two licences. Three months later, in July, the new coalition announced that it was suspending the granting of new licences for oil exploration, pointing out that Greenland was committed to developing strategies for renewable energy, promoting hydropower projects, and contributing to efforts that aim to tackle global climate change.

Egede emphasised this was the path forward at the UNFCCC COP26[5] meeting in Glasgow in November, when he declared that Greenland would join the 2015 Paris Agreement and that the country had ambition to emerge as an internationally recognised leader in green energy. At COP26, Greenland joined ten other national or subnational governments to form the Beyond Oil and Gas Alliance (BOGA), which is committed to deliver a transition away from oil and gas production. Oil exploration and Greenland Minerals aside, other mining companies with interests in Greenland were soon reassured by the new government that the country is still open for business and welcomes bids for prospecting and exploration licences, but not for uranium. Although uranium extraction and oil exploration no longer appear to be promising areas for hopeful investors in Greenland's resource sector (Christiansen ibid.), the subsurface nonetheless remains at the heart of much of Greenland's domestic politics and discourses on political and economic independence. To emphasise this, in September 2021, GEUS carried out a seabed survey commissioned by De Beers off Greenland's west coast near Maniitsoq as a first phase in determining the presence of marine diamond deposits.

The new government's affirmation of Greenland's commitment to developing a mining industry was underscored when, also in July 2021, ICC-Greenland joined the Arctic Economic Forum (AEC), separately from its parent organisation (ICC was one of the founding members of the AEC in 2014), with the intent to promote Greenland's pro-mining stance and to argue that mineral extraction, along with tourism, was essential for diversifying the economy. Greenland's move to accede to the Paris Agreement as a signatory in its own right means that it does not wish to be bound by Denmark's commitments to the treaty, especially as restrictions on CO_2 emissions would hinder the development of mines. As Naaja Nathanielsen, Greenland's Minister of Finance, Minerals, Justice and Gender Equality, put it in February 2022, "The future of Greenland's exploration and mining industry looks bright and evolves with the ever-growing understanding of the geology, which constantly creates new exploration targets." She went on to describe how:

> The country's robust combination of attractive geology, and its potential for high-grade, high-volume green mineral deposits, is further supported by its geographic diversity. Many deposits are located within reasonable proximity to settlements that host airports, electricity, and other services. Others are located close to coastal waterways, enabling the bulk transport of minerals to global markets.[6]

In July 2021, Greenland also joined the European Raw Materials Alliance (ERMA). This was launched in September 2020 as part of a European Commission Action Plan, which aims to reduce Europe's dependency on third party countries for its raw materials as it moves towards being carbon-neutral by 2050. In particular, relying exclusively on China is not a preferred option, a view expressed at the ERMA's launch.[7] The European Union has identified Greenland as a major supplier for most of the critical minerals it needs, including its rare earths, and it is perhaps no coincidence that the European Commission is establishing an office in Nuuk.

As attention has turned to thinking about Greenland as a rare earths frontier, geo-securing critical minerals is strategically important as competition intensifies and global demand increases. And just as Vakulchuck and Overland (2021) point out that China's influence in the critical minerals sector in Central Asia is of serious concern to US security considerations, the US has been similarly anxious about potential Chinese influence in Greenland. American interest in Greenland has extended recently with the signing of agreements that set out a framework for co-operation on mineral sector governance, technical engagement, and capacity-building as well as with the opening of a consulate in Nuuk in June 2020. It is worth pointing out that the US had originally maintained a consulate there from 1940 to 1953 (see Chapter 4). As I mentioned in Chapter 1, it was established following the German invasion of Denmark to provide direct contact and representation between the US government and Danish officials and was eventually closed because of an economy drive in the US diplomatic service (Fisher 1954). When the current consulate opened, a major difference between it and its wartime and early Cold War period antecedent was emphasised: that its purpose is to liaise directly with the Greenlandic government (Cully 2021).

An exploitation licence has been granted to the Australian company Ironbark to develop a lead-zinc mine in Citronen Fjord in Peary Land in the country's far north, and recently to Tanbreez, another Australian company, which is developing a rare earths project in South Greenland. A large ilmenite (an iron-titanium oxide) mine near the depopulated and closed settlement of Moriusaq in Northwest Greenland's Avanersuaq region has also been approved. Exploitation will see ilmenite extracted from coastal sand by Dundas Titanium, a company registered in Greenland and 100% owned by London-based Bluejay Mining. Alba Mineral Resources has a similar project application under review for ilmenite extraction along the same coast and is also scoping out the prospects for iron ore in the Qimusseriarsuaq/Melville Bay region. It is of interest to note that sand and gravel are being highlighted as providing an opportunity for Greenland to become a major exporter of aggregates and help meet increasing global demand (Bendixen et al. 2019). In this way, Greenland's ambitions connect the granular to the global, giving another dimension to how it is implicated in an earthly geopolitics.

Several other mining projects around the country are in various stages of planning and exploration, including at sites north of Qaanaaq, such as Inglefield Land (for copper, gold, cobalt, and nickel), in Melville Bay (again, for iron ore), the southern part of the Upernavik district (for nickel, copper, and platinum group metals), and a number of others on the east coast. Amaroq Minerals is also planning to bring the historical Nalunaq gold mine back into production, while Black Angel Mining has a similar plan for the Black Angel Mine at Maarmorilik, a lead-zinc extractive venture which was operated by Danish mining company Greenex between 1973 and 1990.

Re-making Nuuk

Attracting investors is a form of extraction – in terms of getting them to commit the resources needed to finance the mining industry as well as other economic ventures. Greenland's growing tourism sector, which is an extractive industry in itself, plays with images of a country defined by untouched nature and wildness, a place where adventure awaits the intrepid visitor. It is marketed as remote and distant, yet easily reached by convenient direct air routes via Copenhagen and Iceland. But for politicians, planners, and architects, Greenland remains a technical problem that still needs to be solved, just as the GTO thought it to be in the 1950s and 1960s. Geographical distances between towns and settlements are considerable, presenting a significant challenge for transport routes and telecommunications. There is limited housing, especially to meet demand for Nuuk's growing population, and overcrowding in poor quality dwellings in other towns and in the smaller settlements is a serious social and health matter. There is little infrastructure for mines, whether in the form of harbours and airstrips, and while tourism is promoted heavily, there are few facilities, especially in terms of accommodation, for visitors outside larger towns such as Nuuk and Ilulissat.

Someone visiting Nuuk for the first time would quickly get an impression that Greenland's capital is changing rapidly. In this bustling city of almost 19,000 people, which is almost one-third of the country's total population, new buildings are being erected and new suburbs constructed, a new harbour is continuing to grow, and the airport is being extended with the aim of being able to accommodate transatlantic airliners. This is where the anticipatory politics concerning the future of a global Greenland in a new, global Arctic play out (Dodds and Nuttall 2019). As I mentioned at the beginning of this chapter, to talk with politicians, extractive industry employees, Greenland-based consultants, and business people visiting Nuuk, it is often difficult to get away from ideas about exciting opportunities on a new 'resource frontier,' which require further political and business development.

Nuuk is also where securitisation, geo-economics, geophysics, urbanisation, and environmental change as well as self-determination and geo-security interact and are felt in a number of material, sensorial and embodied ways, that reflect history and worldmaking (Dodds and Nuttall ibid.). To walk around the city is to have an encounter with the traces of its colonial past and Greenland's earlier enclosure for the purposes of the trade in living marine and terrestrial resources and to glimpse a vision of the future. Material remains of the institutions and machinery of trade and maritime networks that were put in place in the eighteenth century are evident in the Kolonihavn area. *Kolonihavn* is Danish for 'the colony harbour,' but it comprises part of a larger area known in Greenlandic as Nuutoqaq, or 'old Nuuk.' This is where the KGH established its trade headquarters at Godthåb and where the city's oldest houses and buildings can be found as well as the old harbour of Sissiugaq. Greenland's national museum is housed in an old KGH warehouse, built in 1936, along with blubber stores, coopers' sheds, and other buildings, all of them memorials to the resource extraction that shaped colonial Greenland. The smell of tar, whale oil, and sealskins still lingers in these old buildings.

Figure 3.1 Qinngorput is one of the newer suburbs in Nuuk and is a key project of the city's housing strategy (Photograph Mark Nuttall).

Hans Egede House, which dates from 1728 when Egede moved from Habets Ø and is the oldest house in Greenland, is at Kolonihavn too. Until a few years ago, it was the residence of the Greenland premier and is used for official government receptions and functions – the premier's current residence is a few houses away, along the same road and nearer the museum. Close by, a statue of Egede stands near Nuuk's Lutheran cathedral (*Annaassisitta Oqaluffia*), which was built in 1848–1849. In summer 2020, the statue, which was erected in 1922, caught wider global public attention during the Black Lives Matter protests that took place in many countries. On Greenland's national day on 21st June, Egede's statue was daubed with red paint to represent blood and tagged with 'Decolonize' written in English. Nine days later, a duplicate statue outside Frederik's Church in Copenhagen was vandalised in a similar way. The statue in Nuuk had been defaced by paint on several occasions in the 1970s and again in 2012 and 2015, but this most recent act provoked anxious and angry debate, expressed vociferously on social media, about the statue as a symbol of oppression, glorifying colonisation and the enslavement of the Greenlandic people. In response, Sermersooq municipality – the administrative region of which Nuuk is a part – ran an online poll to determine whether the statue should be removed. A majority who participated in the poll voted in favour for it to remain standing.

On the other side of Nuuk, there is a busy harbour where cargo vessels, oil supply tankers, trawlers, smaller fishing boats, the coastal ferry and, in summer, cruise ships, dock. The port of Nuuk is Greenland's largest and the ownership and operation of all facilities has been transferred from the Government of Greenland to Sikuki Nuuk Harbour, the port authorities. The port has a large container terminal that was built between 2015 and 2017 and which is operated by Royal Arctic Line. It is a distribution hub for the essential supplies that come to Greenland from many parts of the world via Denmark. Danish naval patrol vessels, which are part of the Kingdom of Denmark's Arctic Command fleet, are also stationed at the harbour. Arctic Command's headquarters are located close by, and over the last decade, there has been a greater presence of military personnel in the capital as well as regular military exercises in the Nuup Kangerlua area. Nuuk's residents are often witness to the activities that result from greater investment in the securitisation of Greenland by Denmark as it asserts its position as an Arctic coastal state.

City planners imagine Nuuk as an urban, globally connected, Arctic metropolis. In 2000, a scenario planning process resulted in a report for the municipality. It set out an ambitious plan for Nuuk up to 2050 as an Arctic capital. It contains fascinating maps that envision the development of a large municipal area – beyond the growth of Nuuk itself, hundreds of kilometres of roads snake to an Atlantic airport, hotels, and art museums, ski districts, restaurants on mountain tops similar to what are found in the Swiss Alps, casinos, scientific research centres close to the inland ice, hiking areas with campsites, and a road connecting Nuuk to Kapisillit (which is 75 kilometres northeast of Nuuk). In this vision of the future, roads will stretch down the coast to new communities and inland to mines, base camps for mountaineers, and to the research centres that will be focusing on geology, mining, ecology, climate change, and environmental management.[8] "This will be quite the city twenty years from now," a city planner told me in 2012.

The new suburbs are already spreading around headlands and inlets, as scenarios for growth suggest the capital could have a population of 30,000 within the next couple of decades. It is anticipated that many of these new residents will move to Nuuk from communities such as those in the Upernavik area and from what are considered other 'outlying' districts in North, South, and East Greenland. In discussions I have had with city planners, the view is that Nuuk needs to grow to provide opportunities for young people so that they do not leave Greenland and move to Denmark. Internal movement between small communities and the capital as well as other towns has always gone on, of course, and the reasons for this are complex and multifaceted. People make choices about moving to Nuuk and other large towns, and even on to Denmark, because of employment prospects and education opportunities, for instance. But there is a greater trend of the depopulation of settlements as livelihoods based on hunting and fishing become increasingly precarious for many as a result of a changing climate as well as providing fewer economic possibilities.

In the settlements in Upernavik district where I do much of my research, the population is ageing while younger people are moving to Ilulissat or Nuuk. In Kangersuatsiaq, for example, which was once a place renowned for its rich hunting and fishing areas and was considered a model for the sustainability of small settlements, around one third of houses stand empty as young people (and entire families) have moved to Upernavik town, to other villages in the district, or on to Nuuk and other towns. The closure of the fish processing plant in 2011 was one catalyst for this movement away, but the effects of a neglect by govern-ment to invest in Kangersuatsiaq and places like it are also cited by people I know who have left (Nuttall 2022). In June 2019, one young married couple I know who are originally from Kangersuatsiaq but who had moved to Upernavik in 2012, told me over dinner at their home that they were moving to Nuuk. They spoke of their decision as being a difficult one – there is something peaceful about living in Upernavik, they said, but they wanted regular jobs. They felt that, if they stayed, there was no future for them as a family as they could no longer get by on fishing, and they wanted their two children to have education opportunities that were not possible in Upernavik. The sustainability of small communities is also a challenge for municipal authorities and for Self-Rule government departments. Goods and supplies are expensive to ship by container from Denmark and on to the more sparsely populated areas of Greenland on the east, south and northern coasts. Life in the small settlements is, in many ways, possible because of government subsidies, but these are gradually being withdrawn. To make one Greenlandic nation, there are many politicians and planners who favour the depopulation of what they see as these remote areas, and the concentration of people in a few larger towns. To a great extent, Greenlandic politics and the shaping of future Greenlandic society are, to draw on Nick Clark's (2017: 213) phrasing, "implicated with specific geological formations." Potential and possibility emerge from the earth's subterranean depths. Extracted from remote places, it is minerals that will contribute to making the future possible, not hunting and fishing.

To become something new, though, Nuuk first has to unbecome what it is and what it has been. To unbecome involves the erasure of the past, as occupied sites and inhabited buildings transition to decay, dereliction, ruin, construction sites, and regeneration (Fraser 2018). The demolition of the ageing apartment blocks and low-income housing constructed rapidly and cheaply in the 1960s has become central to aspirations for regeneration. In the process, Nuuk has become a site for urban renewal and speculation in real estate. Tim Edensor (2020) argues that cities are transformed, replenished, and regenerated by four processes that are interlinked – there is quarrying, demolition, disposal, and the remediation of landfill sites. Stone has to be supplied, old buildings have to be demolished, and new ones have to be built. Stone and rubble have to be taken away somewhere, and the quarries and landfill sites have to be remediated, especially as they of-ten contain hazardous material. All of this is evident in the capital city. Emma Fraser's account of unbecoming in Detroit is particularly germane to how this process and transformation and replenishment plays out in Nuuk. She argues that unbecoming is a "condition of constantly shifting place transformations" that are connected to demolition and redevelopment (Fraser ibid.: 446). And so to rebuild the centre of Nuuk and to construct new suburbs, there must first be destruction, the accumulation of rubble, and the circulation of pervasive dust.

On a particularly warm day in July 2012, I asked a friend who is originally from South Greenland but who has lived in Nuuk for many years to describe how she felt about the city. In one word: "Dusty," she said. She then went on to say: "Thank goodness for winter, when snow covers the ground and makes it hard for the dust to be blown about." She was not the first person I had heard talking about the city in this way. Indeed, being outside in Nuuk on a windy summer's day often appears to be inviting an inescapable encounter with dust. We were sitting at a window table in a downtown café, having lunch and watching trucks rumble past, carrying rocks from a construction site where new apartments were being erected. Later that afternoon, we walked along the road to observe the ongoing demolition of Blok P and watched the excavators ram their swing arms into the concrete walls. As it does for many other things in the world, Kalaallisut has a rich vocabulary to describe dust and dustiness. In this sense, the word my friend used to talk about Nuuk as not just being dusty but looking dusty was *qaserpoq*. The root alludes to small grains or speckles. *Qaserpoq* has another meaning, indicating that something has the colour of dust. It is also a word for grey. *Qasersarpaa* means to make a person or something dusty. *Qasernerit* means to have dust on it. To dust something or to brush dust away is to *qasiiarpaa* it.

This way of thinking about dust being one defining feature of Nuuk suggests a feeling that it is ephemeral, despite its materiality as fine particles of something solid. Yet this granular material has a pervasiveness that cannot be avoided. It is in the air, on the ground, on window ledges, and in people's houses. It drifts around, falls slowly, and gathers. It gets on your face, in your eyes and in your hair, and on your clothes. It is something that is breathed in. It spots and speckles. As an aside observation, a spotted or speckled seal is *qasigiaq*. Many things can *qasersarpaa* Nuuk – traffic, construction, demolition. There are other words for dust, such as *pujoralak* and *pujualak*. *Sanik*, for instance, means a grain of dust; *sequnneq* is the dust that comes from substances like coal or the smudges on an oil lamp. *Sammukarpoq* means to make something dirty with dust. While we watched more of Blok P being turned to rubble, *pujualak* seemed an appropriate way to describe what was billowing from it and convey the sense that the resulting dustiness was cloudy like smoke; *pujualanerit* means that things are covered by such cloudy dust as it gets blown about, swirls, and settles.

I too have often thought that dustiness is an apt word that captures many of my experiences of being in Nuuk. Beyond that, it also seems an appropriate way of describing encounters with other parts of Greenland. One may think of cold air and fierce winds and, perhaps above all, ice and snow, as defining characteristics of Greenland as Arctic territory. However, the colour of Greenland is not just white in winter or greens and browns in summer, but grey dust, the muddy colour of glacial rivers, and dirty icebergs as well as the dark patches that increasingly dot the inland ice in the form of cryoconite and black carbon deposition (to which I return in Chapter 7). In summer, dust makes itself felt upon a passenger's arrival at the airports in Kangerlussuaq and Narsarsuaq. Runways elsewhere along the coast are dusty as well. Once, in summer 1994, I landed at Mesters Vig, a Danish military outpost on the east coast, after a Twin Otter flight from Akureyri in Iceland.

I was heading with seven others for project work in the Northeast Greenland National Park. My first taste of that part of the country as I disembarked the aircraft was not the rarefied air I had expected would fill my lungs, but a mouthful of dust and grit in my eyes as the wind whipped along the airstrip.

I have had the same experience landing at Thule Air Base, Qaanaaq, and Upernavik. Dust rushes to be the first to greet you even in the High Arctic. Roads are dusty, especially gravel ones. Construction sites are dusty. Demolition sites are dusty. When the snow melts in spring and early summer, North Greenland's settlements become dusty again as sandy, gravel pathways are revealed. When glaciers recede, they leave behind debris, gravel, sediment, sand, and rock flour, fine material that is blown around by wind. A walk along a moraine is to sink in ground that is not firm and it leaves dust on one's boots, but it also gets through them and through one's socks, making your feet and toes feel grainy. One gets dusty after a day visiting people, whether carrying out interviews or socialising, going to a kaffemik (a social occasion with coffee, cake, and food, depending on whether it is a birthday), walking between houses, waiting for the bus in downtown Nuuk. To be in Greenland is to be with dust.

Dust is also one of the defining characteristics of Greenlandic landscapes of development, extraction, construction, and demolition. However, it is common for project proponents to argue that dust produced in mining operations will be minimal and will not pose a risk. When I followed the discussions and public hearings for the Greenland Minerals project at Kuannersuit, for example, dust was central to many of the proceedings. Dust is described in detail in the EIA, which argued that the dust impact of the project would be within the limit values laid down in Greenland's Minerals Act and would not affect humans, animals, and the environment. The EIA presented data from models that the proponents argue show dust deposition would extend less than a few hundred metres from the mine's open pit and transport routes. The EIA also argues that emissions of dust and gaseous air pollutants during the construction phase would mainly be expected from the blasting and excavation at the new port, the pit site, main access road, pipelines, and process plants, and emissions from ships, mobile equipment, and power generation. It concludes that dust emissions from construction activities will be local and temporary. Local stakeholders, such as Greenland's association of municipalities and environmental and conservation groups, disagreed, arguing that the Kuannersuit project would increase the amount of dust in Narsaq and Qaqortoq and that the EIA was an insufficient assessment of dust and particle emissions.

Dust in the air and upon the ground can be the result of digging out rock, mining, and crushing ore, excavating sand deposits, and demolishing a building, but it can also be the result of constructing one. Caroline Steedman writes that dust "performs an action of perfect circularity" (Steedman 2001: 160) – 'to dust' means something can be removed, but it can also be put there. In Greenland, dust was produced in the 1950s and 1960s by developing towns to house people from the small settlements that were closed (and which were left to become ruins with their own dustiness), so implying modernisation, progress, and the design of a new society and economy, while at the same time ending a way of life. As the buildings aged, asbestos and damp made their presence felt in people's lungs and bones. Now, dust is an inevitable but essential by-product of self-government, nation-building, and state-formation as Greenland is forged, regenerated, and new forms of society and economy are designed, while the buildings that have stood as edifices of a previous established order erected in the 1950s and 1960s are erased. Greenland's architectural experiments from that period are ending in rubble and dust just as the blasting of foundations for new structures creates dust that the future is being built upon. Greenland's extractive future also begins with chipping away on rock with a geologist's hammer or blasting it for minerals with each new mine that is scoped out, approved, and opened and when the diggers get to work to unearth the resources that have become something valuable. However, dust settles after explosions and demolition and it remains in the form of granular fragments that serve as a reminder of what has been destroyed and removed, whether they are rock outcrops, hills, or buildings. As Caroline Kopf (2020) writes with reference to the daily encounters railway workers in Senegal have with the granular materiality of toxic dust, it is to be aware of the effects that the mobility and circulation of materials have on bodies and places.

For several years, after Blok P had been completely demolished and all the rubble had been taken away from the site by truck in 2012, the area it had covered remained open and unfilled. For a while, part of it was used as a children's playground. For some people I know in Nuuk, including those who had once lived in Blok P, its absence – indeed its spectral nature – signified its enduring presence. Memories of living there, and what Blok P represented, were not obliterated and carried away with the dust and the debris. Even now, as a new school is built on the site Blok P once occupied, it is almost as if, as Sarah Surface-Evans (2020: 154) puts it, "the 'sense' of haunting prevents erasure." But Blok P itself represented a hazard in which dust from asbestos and other toxins were present in the walls, ceilings, floors, stairwells, and which seeped into the bodies of its human residents.

Surface-Evans considers how cityscapes are haunted landscapes and evokes Maria Tumarkin's (2005) definition of traumascapes as places that have been marked by traumatic events (Surface-Evans ibid.: 153). While Tumarkin discusses places that have become associated with traumatic death or acts of terrorism, Surface-Evans argues that traumascapes can encompass a diversity of places – the common theme is that they are associated with traumatic events that come to characterise them. For some, Nuuk's layout and housing constitute the cityscape as traumascape, memorialising the policies and events of the 1950s and 1960s. Of course, there are many residents who would resist this notion, but the traces of colonisation are evident at Nuutoqaq – even in the aroma of the tar and the blubber barrels at the museum buildings – and the legacy of social and economic change is memorialised in the architecture of those apartment blocks where people were relocated to when settlements were closed down in the 1960s and 1970s. The demolition of Blok P, as well as the demolition of other apartment blocks, also brings a sense of loss. People made lives there and raised families. For some, having to leave Blok P and move to a new dwelling was painful.

Figure 3.2 Blok P in Nuuk, standing empty shortly before demolition began in 2012 (Photograph Mark Nuttall.)

Adam Kaasa (2020: 214) argues that processes of urban erasure, urban demolition, and urban renewal require structures of legitimacy to be enacted and produced through materials. These are most obviously rubble, debris, refuse, dust, and ruins, but he argues that structures of legitimacy also involve different kinds of materials that lead to erasure, the accumulation of rubble, and eventual renewal. Such materials are the necessary and obvious building materials such as concrete, bricks, and steel, but include the reports, arguments, and visualisations of urban planning as well as architectural drawings, blueprints, and the products such as models of new neighbourhoods that are required to bring the built environment into being (Kaasa ibid.: 223).

In addition to mines and hydrocarbon exploration, projects to lengthen runways (and build new passenger terminals) at the airports in Nuuk and Ilulissat to accommodate large international aircraft and boost business and tourism, and build an airport at Qaqortoq, as well as the construction of Siorarsiorfik, another new suburb of Nuuk that is currently in the planning stage, have dominated political and public debate – at least in the capital.[9] The airports and the construction of new residential areas involve the blasting, excavation, quarrying, transport, and crushing of thousands of tons of rock. Blasting for the new airports began during the winter of 2019–2020, with the expansion in Nuuk being especially controversial. Opponents organised local campaigns and favoured a new airport at a place called

Angissunguaq, an island to the southwest of Nuuk, but the government approved the expansion of the existing runway at the domestic airport, which lies close to the city, and the construction of a new terminal building (which is critical for Nuuk's projection of an image as an Arctic metropolis). An airport at Angissunguaq would have allowed for a longer runway, located at a safe distance from populated areas, but the government argued that it would have been a far costlier engineering challenge involving several kilometres of roads, tunnels, and bridges. It is one project in the scenario planning process I discussed above that has not been realised. Reshaping the airport in the city seemed an easier political decision to make.

In October 2019, residents who attended a public meeting with the construction company expressed their concerns over noise, dust, and vibration problems, and with the blasting away and levelling of some of the lower reaches of Quassussuaq (Lille Malene), the mountain below which the existing airport sits. The company's civil engineer tried to reassure them that measures would be put in place to ensure that noise would be minimised. He even pointed out that the explosives used are more environmentally-friendly than initially outlined in the project's EIA. By February 2020, many people were already feeling exhausted by several weeks of blasting. The hours of explosive activity had been increased, the extent of how much of the mountainside was being levelled was becoming apparent, and construction of a roundabout and a new airport road was adding to the noise and disruption. Some took to social media or posted comments on newspaper websites to express their anxieties over the prospect of living with two or more years of noise and dust, to say nothing of the disturbance anticipated once the airport is complete and transatlantic aircraft fly in and out of Nuuk from 2024. The construction has since become something of a spectacle, with videos posted on YouTube that chronicle the largest blasts, the movement of rock and earth, and clouds of billowing dust. Whether it is the underwater sonic boom from seismic surveys in Melville Bay (see Chapter 6), the blasting of explosives in Nuuk for airport construction, or wrecking balls pounding on concrete walls to demolish apartment buildings, noise is an increasingly constant and unruly but necessary accomplice to Greenlandic future-making.

In an influential essay on various dimensions of remoteness, Edwin Ardener (1987) once wrote how remote areas are full of innovators and can experience a sequence of innovations and projects that appears endless. There is always something new to build, another pier, harbour, or housing district, for example, or another quarry to develop, another road to push through, and airports to construct. This is apparent in contemporary Nuuk as a centre for abstraction. But it is also a place where it is possible to reflect on how it is made into an attractive site for innovation by what Baudrillard (1996: 40–41) calls the "boundless possibilities" of "abstract integration." Innovators, whether they are architects, urban designers, fishing company executives, mining engineers, resource managers, or policymakers, are essential to the historical and contemporary processes of spatial, social, and cultural transformation and urban regeneration as well as the materialisation of atmosphere and affect that have invented and continue to reinvent Greenlandic towns, settlements, resource spaces, and the country's remote, wild edges as zones of possibility and opportunity. Anything seems possible in Nuuk. It is a blank slate for design and for reimagining it as a Global Arctic city.

In their work on roads in Peru, Penny Harvey and Hannah Knox talk about the engineer-*bricoleur*, someone who engages with "uneven, unruly, and unstable environments out of which infrastructures are made" (Harvey and Knox 2015: 7). Greenland has few roads. Most are in towns – Nuuk, for example, has around 100 kilometres of roads within its city limits, which is a considerably large part of the country's network – while others were first constructed by the US when they built their air bases during World War II and other initiatives during the Cold War (see Chapter 4). Roads, such as those stretching around Thule Air Base and Søndre Strømfjord in arterial patterns, were built to move personnel and equipment from harbours to and around the main base areas, to allow vehicles to service aircraft, to connect the main base sites to outlying radar, radio, and weather stations, and to reach rocket launch sites. Roads were also built from these bases to the edges of the inland ice, which became central to US military plans for protection and conceal-ment. Connecting Greenland's towns and settlements by a road network would be a considerable engineering and logistical challenge, so the air transport network is likely to be expanded once the new airports are operational. For Harvey and Knox (ibid.: 22), roads have a "capacity to conjure a sense of the potential of enhanced connectivity." This has "enabled the consolidation of singular sovereign national territories" and contributed to "the production of vibrant transactional networks." Roads are connective technologies that hold promise "for both economic develop-ment and political integration." In the same way as they write about roads in Peru as state space, I see infrastructure projects such as airports and harbours in Green-land as spaces that are intended to make state formation possible.

Even as Greenlandic politicians and business interests imagine a Global Green-land, viewed from Denmark, other parts of Europe, and North America, Nuuk too is a place on the edge – an Arctic frontier for construction companies and architects who see possibilities for lucrative contracts and for the hope of being celebrated globally for award-winning design in the outer reaches of the Danish Realm. One recent example of this is Anstalten Correctional Facility, the new prison in Nuuk which was completed in 2019. This description appears on the homepage of the Danish architects who designed it:

> Nestled in the rugged terrain of Greenland's seaside capital, Anstalten in Nuuk - a new correctional facility, is the setting for progressive rehabilita-tion, and a bold statement about the power of architecture to affect human behavior. Openness, light, views, security and flexibility are the leading val-ues behind the design of the first such facility in the capital of Greenland. The project matches the unique and beautiful surroundings and supports the focus of the Danish Prison Service on both punishment and rehabilitation.[10]

The company's website goes on to describe how, architecturally, the prison is "composed of accurately shaped blocks, which in their positioning follow the con-tours of the rocky landscape" while "burnt sienna corten steel facades, impacted by the local climactic (sic) conditions, complement the surrounding landscape." Openness is a central theme – large windows not only let in light, but "draw nature inside." Overall, the building is designed and situated to "appear subordinate to its surroundings."

Such description brings to mind Baudrillard's notion of atmospheric values in accounting for the design of buildings. The materials used to construct the prison, and which are intended to support both punishment and rehabilitation, blend with the land, sea, and sky. Concrete, steel, wood, and glass draw their meaning, as the architects and designers see it, from the Arctic environment. But, in the sense that Baudrillard suggests, they have an abstractness that provides for the possibility of "a universal play of associations among materials, and hence too a transcendence of the formal antithesis between natural and artificial materials" (Baudrillard ibid.: 39). Building materials as well as colours, volume, space, light, views towards the sea, and walls decorated with animal motifs from Greenlandic stories and oral history intermingle and are used as atmospheric elements. Similarly, the Danish company MT Højgaard International, which won the contract for the construction of the new airport buildings in Nuuk, has imagined the terminal building, with steel structures and sections of glass facades, to reflect the mountain landscape that surrounds the city.

Nuuk represents the possibilities of innovation, extraction, renewal, progress, and sustainability. While its historical roots are deep in colonial times, Nuuk is being refreshed, being made anew as it undergoes a process of unbecoming – shedding its colonial and post-colonial structures and appearance – and becoming something other in a process of regeneration. The colonial and post-colonial traces and rotting structures must first be extracted. Dust, which can be a source of irritation, grime, and pollution, must be produced and mobilised in order for the past to be dislodged and removed. Wright et al. (2013) discuss how dust blurs the line between order and disorder – it is a nuisance but also shapes and reshapes human adaptive responses to changes in ecosystems. Michael Marder (2016) sees dust as mundane, yet pervasive. In its accumulation in layers, he writes that it is a gathering place that blurs boundaries between the living and the dead, between the present and the past. People have memories of the construction that happened in Nuuk in the 1960s and 1970s as turning the town into a place of dust. And while dust arises from disintegration it also signifies renewal (Marder ibid.). So while mining may be seen as one critical pillar of Greenland's economic development policy and while some politicians remain hopeful that oil may still flow one day, despite current opposition and official policy, the building of transport infrastructure, such as airports and harbours, and the design of fresh urban spaces for a new, emergent Greenland, whose people from its outer edges are compressed into an ever-growing capital, are no less predicated on the extraction, removal, levelling, crushing, fashioning, and shaping of ancient rock, and on digging, unearthing, blasting, and demolition.

Notes

1 Ove Karl Berthelsen, "Foreword" in Ministry of Industry and Mineral Resources (2009), *White Paper on the status and development of the aluminium project*, Nuuk: Government of Greenland, p. 5.

2 Richard Mills, "Greenland offers exploration homerun potential." Mining.com, 10 August 2013.
3 Kalantzakos, S. 2018. *China and the Geopolitics of Rare Earths*. Oxford: Oxford University Press.
4 Nakano, J. 2021. *The Geopolitics of Critical Minerals Supply Chains*. Washington, DC: Center for Strategic and International Studies.
5 UNFCCC: United Nations Framework Convention on Climate Change; COP means Convention of the Parties to the UNFCCC.
6 Naaja H. Nathanielsen. 2022. "Exploring Greenland's critical mineral potential." *Innovation News Network*, 21 February 2022, https://www.innovationnewsnetwork.com/exploring-greenlands-critical-mineral-potential/18566/.
7 European Commission. (2020). ERMA Launch (European Raw Materials Alliance) https://webcast.ec.europa.eu/erma-launch-european-raw-materials-alliance.
8 *Nuuk/En Arktisk Hovestad – klar til et nyt åhundrede.* A scenario for the period 2000–2050 produced by Tegnestuen Nuuk A/S in January 2000.
9 Some of the ideas that are elaborated here in the closing section of this chapter were first sketched out in Mark Nuttall (2022), "Wild lands, remote edges: formations and abstractions in Greenland's resource zones," in Arthur Mason (ed.) *Arctic Abstractive Industry*. Oxford: Berghahn, pp. 83–107.
10 Schmidt Hammer Lassen Architects, *https://www.shl.dk/anstalten-correctional-facility/*. Accessed 15 March 2021.

References

Arboleda, Martín. 2020. *Planetary Mine: Territories of extraction under late capitalism*. London and New York: Verso.

Ardener, Edwin. 1987. "Remote areas: Some theoretical considerations." In Anthony Jackson (ed.) *Anthropology at Home*. London and New York: Tavistock, pp. 38–54.

Ayeh, Diana. 2021. *Spaces of Responsibility: Negotiating industrial gold mining in Burkina Faso*. Berlin and Boston: Walter de Gruyter.

Baudrillard, Jean. 1990. *Seduction*. New York: St Martin's Press.

Baudrillard, Jean. 1996. *The System of Objects*. London: Verso.

Bendixen, Mette, Irina Overeem, Minik T. Rosing, Anders Anker Bjørk, Kurt H. Kjær, Aart Kroon, Gavin Zeitz and Lars Lønsmann Iversen. 2019. "Promises and perils of sand exploitation in Greenland." *Nature Sustainability* 2: 98–104. https://doi.org/10.1038/s41893-018-0218-6.

Bjørst, Lill Rastad. 2020. "Stories, emotions, partnerships and the quest for stable relationships in the Greenlandic mining sector." *Polar Record* 56, E23. doi: 10.1017/S0032247420000261.

Bridge, Gavin. 2004. "Mapping the bonanza: geographies of mining investment in an era of neoliberal reform." *The Professional Geographer* 56 (3): 406–421.

Carse, Ashley and David Kneas. 2019. "Unbuilt and unfinished: The temporalities of infrastructure." *Environment and Society: Advances in Research* 10: 9–28.

Childs, John. 2020. "Extraction in four dimensions: Time, space and the emerging geo(-)politics of deep-sea mining." *Geopolitics* 25 (1): 189–213.

Christiansen, Flemming G. 2021. "Greenland petroleum exploration history: Rise and fall, learnings, and future perspectives." *Resources Policy* 74, 102425. https://doi.org/10.1016/j.resourpol.2021.102425.

Clark, Nick. 2017. "Politics of strata." *Theory, Culture and Society* 34(2–3): 211–231.

Cons, Jason and Michael Eilenberg. 2019. "Introduction: On the new politics of margins in Asia: Mapping frontier assemblages." In Jason Cons and Michael Eilenberg (eds.)

Frontier Assemblages: The emergent politics of resource frontiers in Asia. Chichester: John Wiley and Sons Ltd.

Cully, Eavan. 2021. "Setting up shop in Nuuk." *The Foreign Service Journal* May 2021, pp. 34–37.

De Gregori, Thomas R. 1987. "Resources and not; they become: An institutional theory." *Journal of Economic Issues* 21 (3): 1241–1263.

De Rijke, Kim. 2017. "Produced water, money water, living water: Anthropological perspectives on water and fracking." *WIREs Water* 5 (2): https://doi.org/10.1002/wat2.1272

Dodds, Klaus and Mark Nuttall. 2016. *The Scramble for the Poles: The geopolitics of the Arctic and Antarctic*. Cambridge: Polity.

Dodds, Klaus and Mark Nuttall. 2018. "Materialising Greenland within a critical Arctic geopolitics." In Kristian Søby Kristensen and Jon Rahbek-Clemmensen (eds.) *Greenland and the International Relations of a Changing Arctic: Postcolonial paradiplomacy between high and low politics*. London and New York: Routledge, pp. 139–154.

Dodds, Klaus and Mark Nuttall. 2019. "Geo-assembling narratives of sustainability in Greenland." In Ulrik Pram Gad and Jeppe Strandsbjerg (eds.) *The Politics of Sustainability in the Arctic: Reconfiguring identity, space and time*, London and New York: Routledge, pp. 224–241.

Edensor, Tim. 2020. *Stone: Stories of urban materiality*. Cham: Palgrave MacMillan.

Fisher, Wayne W. 1954. "The passing of Godthaab." *Foreign Service Journal* 31 (3): 30–31, 52.

Fraser, Emma. 2018. "Unbecoming place: Urban imaginaries in transition in Detroit." *Cultural Geographies* 25 (3): 441–458.

Government of Greenland. 2014. *Our Mineral Resources – Creating Prosperity for Greenland: Greenland's Oil and Mineral Strategy 2014–2018*. Nuuk: Ministry of Labour, Industry and Trade.

Government of Greenland. 2020. *Greenland's Mineral Strategy 2020–2024*. Nuuk: The Ministry of Mineral Resources. February 2020, https://govmin.gl/wp-content/uploads/2020/03/Greenlands_Mineral_Strategy_2020-2024.pdf.

Government of Iceland. 2020. *Greenland and Iceland in the New Arctic*. Reykjavik: The Ministry for Foreign Affairs.

Harvey, Penny and Hannah Knox. 2015. *Roads: An anthropology of infrastructure and expertise*. Ithaca: Cornell University Press.

Igoe, Jim. 2017. *The Nature of Spectacle: On images, money, and conserving capitalism*. Tucson: University of Arizona Press.

Johnson, Leigh. 2010. "The fearful symmetry of Arctic climate change: Accumulation by degradation." *Environment and Planning D* 28 (5): 828–847.

Kaasa, Adam. 2020. "The matter of erasure: Making room for Utopia at Nonoalco-Tlatlelolco, Mexico City." In Philipp Schorch, Martin Saxer and Marlin Elders (eds.) *Exploring Materiality and Connectivity in Anthropology and Beyond*. London: UCL Press, pp. 214–227.

Klinger, Julie. 2017. *Rare Earth Frontiers: From terrestrial subsoils to lunar landscapes*. Ithaca and London: Cornell University Press.

Kopf, Charline. 2020. "The dynamics of toxic dust." *Anthropology Today* 36 (6): 17–20.

Latour, Bruno. 1987. *Science in Action: How to follow scientists and engineers through society*. Cambridge, MA: Harvard University Press.

Luning, Sabine. 2018. "Mining temporalities: Future perspectives." *The Extractive Industries and Society* 5 (2): 281–286.

Marder, Michael. 2016. *Dust: Object lessons*. London: Bloomsbury Academic.

Mason, Arthur. 2022. *Arctic Abstractive Industry: Assembling the vulnerable and valuable North*. Oxford and New York: Berghahn.

Mezzadra, Sandro and Brett Neilson. 2017. "On the multiple frontiers of extraction: Excavating contemporary capitalism." *Cultural Studies* 31 (2–3): 185–204.

Nuttall, Mark. 2010. *Pipeline Dreams: People, environment and the Arctic energy frontier.* Copenhagen: IWGIA.

Nuttall, Mark. 2013. Zero tolerance, uranium and Greenland's mining future. *The Polar Journal* 3 (2): 368–383.

Nuttall, Mark. 2017. *Climate, Society and Subsurface Politics in Greenland: Under the Great Ice.* London and New York: Routledge.

Nuttall, Mark. 2022. "Places of memory, anticipation, and agitation in Northwest Greenland." In Kenneth L. Pratt and Scott A. Heyes (eds.) *Memory and Landscape: Indigenous responses to a changing North.* Athabasca: Athabasca University Press, pp. 157–177.

Phillips, Adam. 1994. *On Flirtation.* Cambridge, MA: Harvard University Press.

Richardson, Tanya and Gisa Weszkalnys. 2014. "Introduction: Resource materialities." *Anthropological Quarterly* 87 (1): 5–30.

Schama, Simon. 1995. *Landscape and Memory.* London: Harper Collins.

Stammler, Florian and Aitalina Ivanova. 2016. "Resources, rights and communities: Extractive mega-projects and local people in the Russian Arctic." *Europe-Asia Studies* 68 (7): 1220–1244.

Steedman, Carolyn. 2001. *Dust: The archive and cultural history.* Manchester: Manchester University Press.

Surface-Evans, Sarah. 2020. "Traumascapes: Progress and the erasure of the past." In Sarah Surface-Evans, Amanda E. Garrison and Kisha Supernant (eds.) *Blurring Timescapes, Subverting Erasure: Remembering ghosts on the margins of history.* Oxford and New York: Berghahn, pp. 149–170.

Takeyama, Akiko. 2016. *Staged Seduction: Selling dreams in a Tokyo host club.* Redwood City: Stanford University Press.

Tavory, Iddo. 2009. "The structure of flirtation: On the construction of interactional ambiguity." In Norman K. Denzon (ed.) *Studies in Symbolic Interaction.* Bingley: Emerald Group Publishing Limited, pp. 59–74.

Tumarkin, Maria. 2005. *Traumascapes: The power and fate of places transformed by tragedy.* Victoria: Melbourne University Press.

Vakulchuck, Roman and Indra Overland. 2021. "Central Asia is a missing link in analyses of critical materials for clean energy transition." *One Earth* 4 (12): 1678–1692.

Weszkalnys, Gisa. 2011. "Cursed resources, or articulations of economic theory in the Gulf of Guinea." *Economy and Society* 40 (3): 345–372.

Weszkalnys, Gisa. 2015. "Geology, potentiality, speculation: On the indeterminacy of first oil." *Cultural Anthropology* 30 (4): 611–639.

Willow, Anna J. and Sarah Wylie. 2014. "Politics, ecology, and the new anthropology of energy: Exploring the emerging frontiers of hydraulic fracking." *Journal of Political Ecology* 21 (1): 237–257.

Wright, David K., J. Andrew Darling, Barnaby V. Lewis, Craig M. Fertelmes, Chris Loendorf, Leroy Williams and M. Kyle Woodson. 2013. "The anthropology of dust: Community responses to wind-blown sediments within the Middle Gila River Valley." *Human Ecology* 41: 423–435.

Zalik, Anna. 2018. "Mining the seabed, enclosing the area: Proprietary knowledge and the geopolitics of the extractive frontier beyond national jurisdiction." *International Social Science Journal* 68 (229–230): 343–359.

Zimmermann, Erich W. 1951. *World Resources and Industries: A functional appraisal of the availability of agricultural and industrial resources,* 2nd edn. New York: Harper and Brothers.

4   Geo-security and subterranean Greenland

A Cold War legacy

In August 2019, then-US President Donald Trump remarked that he would like to explore the idea of buying Greenland from Denmark. This provoked considerable global media interest, with numerous articles and op-eds promptly appearing in print and online news outlets. It was characteristic of the kind of reporting about the narrative of a scramble for resources and territorial claims in a rapidly changing Arctic that has appeared so often over the last couple of decades (Dodds and Nuttall 2016). Trump's focus on Greenland was explained in these reports and other dispatches in terms of the country being attractive to the US because of its potential mineral wealth and geostrategic location. Greenlandic and Danish leaders – then-Premier Kim Kielsen and Denmark's Prime Minister Mette Frederiksen, foremost among them – were quick to dismiss the idea of an American purchase. They pointed out that Greenland was not for sale, that it enjoyed a considerable degree of political autonomy, and that the Self-Rule Act meant that Greenlanders were free to decide if and when they wished to pursue independence.

Somewhat miffed, Trump reacted by cancelling a planned official visit to Denmark. Yet what stood out in much of the reporting was that journalists and feature writers assumed that many readers – especially in the US – likely knew very little about Greenland and its contemporary politics, economics and social and cultural dimensions, its geopolitical history, and the controversies and realities surrounding resource development and extractive industries. In September 2019, for example, an article in *The Wall Street Journal* felt it necessary to explain that Greenland "is located east of Canada between the Arctic and Atlantic oceans."[1] As Ingo Heidbrink (2022: 9) sees it:

> Public knowledge about Greenland in the US can be summarized as being the largest island on the globe, completely covered by an icesheet, in geological terms belonging to the North American continent and previously the stepping-stone for Vikings when they sailed to America.

Yet in response to Trump's suggestion of buying Greenland, former Premier Kuupik Kleist expressed concern that Greenland was once again at the centre of global power struggles over the Arctic as it was during the Cold War and that

DOI: 10.4324/9781003175421-5

the US government was positioning itself to invest in infrastructure that would secure their military interests in Greenland as a region vital to securitisation of the northern hemisphere.[2]

With Trump out of office a few months later, US interest in Greenland did not dim. In May 2021, Antony Blinken, Secretary of State in President Joe Biden's administration, flew to Kangerlussuaq for a one-day meeting with the new premier of Greenland Múte Egede, the new minister of foreign affairs, trade, climate, and business Pele Broberg, and the Danish Foreign Minister Jeppe Kofod (Blinken had been in Denmark and Iceland prior to his Greenland visit). They toured scientific research facilities, went on a helicopter flight over the edge of the inland ice, heard about the effects of climate change on Greenland, and stood on a ridge viewing musk oxen through binoculars. At the press conference afterwards, journalist Arnaq Nielsen from Greenland's national broadcaster KNR asked Blinken why it was important that the US should prioritise an official visit to Greenland during the early stage of Joe Biden's presidential period. "Tell us specifically," she asked, "why is Greenland so important?"[3] Blinken replied that the US wanted to "to build a true and strong partnership with Greenland." He said that a reason for the visit was:

> ...to demonstrate that the way we see the relationship is as a partnership. We have shared interests; we have shared values. At a time when the world is ever more complicated and challenging, it's very important to reinvigorate ... not only our alliances, but our partnerships with countries that share our interests and values. And that's why we've sought to deepen our engagement here in Greenland.

Jeppe Kofod pointed out that:

> ...the United States is the most important ally to the Kingdom of Denmark – to Greenland, to Denmark. It is the foundation for our security. We share common interests, we share common values, and I believe we have done so over many decades. And this prosperity that we all enjoy is something that is due to the strong transatlantic relationship that we hold so dear. And these values and the world that we are in, we need to protect them, we need to protect our interests, and that is why I'm so happy that Secretary Blinken, Tony, can come to Denmark, to Greenland, and have this close, close cooperation with us. Because we share also the same aspirations for delivering prosperity, economic opportunity, and peaceful cooperation also in the Arctic and North Atlantic.

Jill Franks (2006) has written how islands have a conquerability about them and are thought of as being vulnerable to attack – reasons for why states set about ensuring their securitisation. The US has long been interested in Greenland's strategic location, however, even to the extent of exercising, arguably, a form of *de facto* sovereignty over the territory – or at least large stretches of it – during World War

II and for much of the Cold War (Olesen 2019). US administrations had talked about buying Greenland on two occasions before Trump's declaration of interest, while, as I discussed in Chapter 1, Danish sovereignty over part of Northeast Greenland had been disputed and challenged by Norway in the 1920s. Canada, with Norwegian encouragement and support, had also drafted plans to offer to buy Greenland from Denmark in the early 1930s. Some twenty years before that, however, lobbyists in Canada had argued for a purchase of Greenland and politicians in Ottawa were fearful of American explorer Robert Peary's incursions into Canada's High Arctic islands and were wary of Danish territorial expansion into northern Greenland undermining Canadian sovereignty (Fogelson 1985). These concerns were heightened when Danish-Greenlandic explorer Knud Rasmussen explored part of Ellesmere Island before World War I and later led an expedition by dog sledge from Greenland across Arctic Canada to Alaska in the early 1920s. Understanding this history is key to contemporary debates about Greenlandic self-determination, sovereignty, and security. It is also necessary to know for understanding current American policy towards consolidating bilateral relations not just with the Kingdom of Denmark, but with specific arrangements that are particular to Greenland. These include investment in geophysical research (notably, scientific studies of climate change) and mineral resource investigations, building capacity in education, and strengthening co-operation within the framework of agreements between the US, Denmark, and Greenland around Thule Air Base.

In Chapter 2, I discussed the Danish-enacted modernisation policies of the 1950s and 1960s and their effects on Greenland. As Kristian Nielsen points out, while this transformation of Greenlandic society and economy was underway, a US-led militarisation of northern Greenland was happening concurrently. Greenland remained under Danish sovereignty, and a new society emerged and took shape based on Danish visions for a modern welfare social system, albeit aligned with post-colonial structures of governance. Its northern reaches, however, were subject to a different kind of enclosure and were subsumed as "an integral part of the networked military empire of the US" (Nielsen 2013: 34). In this chapter, I discuss how the militarisation process, which began during World War II and was expanded during the Cold War, placed Greenland at the very centre of geo-security concerns that were not just Arctic-focused, but had a global reach. Central to this were efforts to extract knowledge about Greenland's subterranean nature – specifically what lay within the ice and under the ice, the seabed, and geological formations – but also about its airspace.

Research in the environmental physical sciences was key to the development of this enormous technological system of North American defence and security. Max Weber saw one of the characteristics of bureaucracy as being a concern with areas of jurisdiction that can be ordered by rules and regulations. Documents, plans, and reports are necessary technological elements of this and act as devices of control, power, enclosure, and surveillance. For the US securitisation process to work properly and effectively in northern Greenland, hundreds of studies were needed on the nature of ice, permafrost, the ocean, and the weather. These generated reports that provided the scientific basis for the bureaucratic oversight of military activities and the administration of the region. These studies were carried out by the US military and civilian scientists and engineers – many of them employed at American universities – who were contracted to work on northern Greenland. They resulted in the production of thousands of pages that today constitute a vast archive of reports and manuals. They are the kinds of scientific documents Latour and Woolgar (1986) describe as 'inscriptions' – they stand as representations of northern Greenland that were materialised and given legitimacy by the agencies that produced them. They constitute a body of facts and knowledge about Greenland's inland ice, glaciers, sea ice, its coastal waters, and its geology, mineralisation and its resource potential, as well as its strategic value. Together, they are valuable historical records of how the US enacted research to get to know Greenland and understand its cryospheric nature, but they also had an essential bureaucratic purpose in contributing to the marking out and the enclosure of parts of northern Greenland as US-administered spaces. All this continues to have a legacy today in terms of how the securitisation of Greenland matters to the US. This, in turn, influences bilateral relations between Greenland and the US, and in a wider sense, between the Kingdom of Denmark and the US.

Securing Greenland during World War II

In the 1860s, William Henry Seward, who was Secretary of State in Abraham Lincoln's administration, argued to Congress that the US needed to have both Greenland and Alaska within its national borders so that it could exercise sovereignty and hold power over parts of the North Pacific and North Atlantic, and thus control access to the northern approaches to the North American Arctic (Hough 2013). The US had purchased Alaska from Russia in 1867 for US$7.2 million (equivalent to around US$120 million today). The decision was ridiculed in the American press and the territory was nicknamed Seward's Folly, before the discovery of gold at Sitka in Southeast Alaska in 1872 caused a rethink of the economic possibilities the territory presented. Seward continued to look northwards to Greenland as well as Iceland and he was encouraged by the arguments of Robert J. Walker, an expansionist Democrat politician, who had supported the annexation of Texas in 1845. In 1867, Walker commissioned the US Coast Survey to do a report on Greenland and Iceland, which he delivered to Seward. The report included details on the resources available on the islands. Walker also suggested to Seward that the US should acquire the Danish Caribbean island colonies of St. Thomas and St. John (Heidbrink ibid.: 14–15).

A strategic reason partly inspired the argument for control over these northern regions, but Seward felt that if they could become American territories, there would be significant opportunities presented to the US in the form of greater access to the Arctic for the purposes of exploration, mapping, and ownership of resources in areas that were already claimed by Russia and British Canada, and even extending into the Arctic Ocean. Seward could not gather enough support in Washington DC for a formal approach to be made to the Danes about purchasing Greenland and Iceland as well as their Caribbean possessions – Congress was too worried about the national debt following the American Civil War – but his ideas did mean government support was allocated for further exploration of the High Arctic, which contributed to an understanding of the importance the region had for defence and the potential value of minerals (cf. Fogelson ibid.; Dodds and Nuttall 2018).

In 1881–1884, army officer Lieutenant Adolphus Greely led his second Arctic expedition with 25 men to Lady Franklin Bay on the northern east coast of Ellesmere Island. Promoted by the US Army Signal Corps, its task was to establish a meteorological monitoring station, conduct magnetic and astronomical studies, and carry out exploration along the coast and interior of Ellesmere Island as part a US contribution to the scientific and geographical studies being conducted during the first International Polar Year. However, it was viewed by the US government as the beginning of a new era of American scientific research in the Arctic, and Greely was not only given the task of collecting scientific data, but with breaking the record for reaching the farthest north point by any expedition. And if conditions allowed, Greely was to make an attempt to reach the North Pole. The expedition established Fort Conger as its base and scientific station, but food supplies ran low and relief ships that were to re-provision it failed to reach there because of ice conditions. Starvation and disease followed and only seven men, including Greely, survived. Following this tragedy, the US government ended its financing of Arctic exploration. For the next forty years or so, American expeditions were largely funded by private sponsors and geographical exploration societies (Robinson 2006).

The US recognised Danish sovereignty over the entire island of Greenland when it purchased the Virgin Islands from Denmark in 1917. Greenlanders had no part to play in the negotiations and deliberations that affected the way their country was classified and recognised as Danish territory and how the US considered the island in this diplomatic process (Heinrich 2018). Since the 1850s, Denmark had viewed their colonial possessions in the Caribbean, which supplied sugar, rum, molasses, and tobacco from plantations, as increasingly unprofitable and at various points at the end of the nineteenth century had tried to interest the US and Germany in buying them. The Americans had initially expressed interest in the 1860s, and were eventually enticed to the point of making a deal with Denmark because they saw the purchase of the islands as a strategic move in defence of the Panama Canal (Heibdrink ibid.: 16). They were especially concerned about German naval operations in the Caribbean and Central American waters during World War I, but their recognition of Danish sovereignty was conditional on Denmark not ceding or selling its Arctic territory to any third power (Petersen 2011). Peary, however, protested about the US surrender of any claim over Greenland – he argued that as the island was part of the North American continent, it should be owned by the US. He also drew attention to its economic value because of its rich coal deposits and glacial streams (Fogelson ibid.: 135, n.18).

After Denmark was occupied by Germany on 9 April 1940, Greenland was effectively cut off from the rest of the Danish Realm. Given the circumstances, Henrik Kauffmann, who was Danish ambassador to the US, now considered himself the sole representative of Danish interests in North America. Somewhat reluctantly, Eske Brun and Aksel Svane, the governors of North Greenland and South Greenland (and who were based in Godhavn – now Qeqertarsuaq – and Godthåb, respectively), recognised Kauffmann's position. They invoked a clause in the Greenland Administration Act of 1925 which allowed governors to assume control of their districts in the event of a connection between Denmark and their districts being lost or severed. Brun and Svane reiterated a pledge of allegiance to King Christian X of Denmark, but declared Greenland a self-governing territory. In doing so, they also expressed a hope that the US would nonetheless help keep the Danish flag flying in Greenland if needed. Both Britain and Canada considered the possibility of establishing a presence in Greenland, and Brun and Svane were nervous of a possible Canadian occupation, but the US rejected the idea of any third party intervention, a point it had also reiterated in a declaration in 1920. It was a diplomatic stance that was taken to be an extension of the Monroe Doctrine of 1823. In May 1940, the US sent a Coast Guard cutter to Greenland with supplies and dispatched a consular team to Godthåb to provide direct representation between the US government and Danish officials in Greenland. Canada also sent a consul and vice-consul the following month.

One year after the German occupation of Denmark, Kauffmann signed an 'Agreement relating to the Defense of Greenland' with US Secretary of State Cordell Hull in Washington DC on 9 April 1941. The US did not dispute Danish sovereignty over Greenland, but expressed a feeling of obligation to assist the territory under the Act of Havana of 1940. This provided that European colonies and possessions in the Americas would be controlled and administered by an inter-American force until the end of the war, at which point they would return to their colonial status or become independent if they wished. The US was concerned that Greenland could be occupied by Germany, and so defending it from attack was considered essential to the security of the North American continent. The agreement gave the US the right to construct, maintain, and operate aircraft landing fields, seaplane facilities, and radio and meteorological installations in order to secure Greenland from being converted into a base for strategic aggression by Germany. The defence agreement also gave the US rights to improve and deepen harbours and anchorages, construct roads, communication posts, fortifications, repair and storage facilities, and housing for military personnel.

The Kingdom of Denmark retained sovereignty over any defence areas established in Greenland, but the US government was granted the right of exclusive jurisdiction over them, except in matters relating to those Danish citizens and Greenlanders working or based in any such area. Another main concern for the US, however, was securing the continued production and shipping of cryolite, which was vital to the manufacture of aluminium for military aircraft construction, from the mine at Ivittuut – indeed, even before the 1941 agreement was signed, the US Navy had provided a local defence unit in late 1940 for the mine upon the request of Brun and Svane which allowed the Americans to maintain a neutral position. In 1943, the US Navy established a based at Grønnedal (Kangilinnguit) at the mouth of Arsuk Fjord, east of Ivittuut, to strengthen the defence of the cryolite resource space (Christiansen 2022). Two years before, in 1941, the

US Coast Guard coordinated with Brun to establish the North-East Greenland Sledge Patrol – which later became the Sirius Dog Sled Patrol – to perform journeys of reconnaissance by dog team in winter and spring and by boat in the summer along the northeast coast and monitor potential German activity. The patrol discovered the Germans had established a weather station on Sabine Island, and this was later destroyed by a US air force bomber operating from Iceland (the German personnel had withdrawn before the attack). The Germans also established three other stations, although they were soon seized by the US coast guard.

During World War II, control of the North Atlantic was considered key to military success. Greenland became a strategically vital part of the region, which itself constituted critical sea and air space for the campaigns of the Allied forces. The US built air bases at Narsarsuaq in South Greenland (constructed on a glacial moraine and known as Bluie West One, it became the headquarters for Greenland Base Command), Søndre Strømfjord (Kangerlussuaq) known as Bluie West Eight on the west coast and Ikateq known as Bluie East Two near Ammassalik on the east coast. Ikateq was more of a minor air base and operated from 1942 to 1947, while Narsarsuaq and Kangerlussuaq later became Greenland's main civilian airports. Bluie was the code name for Greenland used by the US military and 14 numbered bases were established. Some were well-constructed airfields, others were basic gravel airstrips, while several were radio and weather stations. The naval facility at Grønnedal was known as Bluie West Seven. Not only was Greenland an essential link on the North Atlantic stepping stone route for naval and merchant fleet convoys and for aircraft movement between North America and Europe, securing the island was crucial for meteorology and for supplying knowledge of weather patterns for sea and air operations. This depended, in turn, on having detailed knowledge of the prevailing weather over a vast area. Understanding what the weather was like in Greenland and the northern North Atlantic at any one time was vital for forecasting what the weather could be like in northwest Europe a few days later (Dodds and Nuttall ibid.).

Greenland was placed at the very centre of an assemblage of military and scientific technology and infrastructure, as the US air bases were constructed and operated by several thousand military personnel and as weather stations monitored atmospheric phenomena. Wallace Hansen's book *Greenland's Icy Fury*, which was published in 1994, is an interesting and well-written account that gives insight into what it was like erecting and then operating a weather/rescue station on Greenland's inland ice on the east coast (Hansen was later a research geologist with the US geological survey and his book also contains wonderful accounts of some of Greenland's geological features). As Klaus Dodds and I have written, Greenland was not viewed as a remote Arctic space in which to enter, explore, traverse, claim and map, but became important to how we look up into the sky and the atmosphere, across the ice, and deep into the sea. The expansion of air routes and the scientific-technological mapping and measurement of northern air, land, and sea spaces and the acquisition and control of meteorological knowledge situated Greenland in an emerging global geopolitical system of security, surveillance, and science (Dodds and Nuttall ibid.: 144).

Securing Greenland during the Cold War

Following the end of the war, Greenland's administration surrendered its emergency powers and Copenhagen reassumed direct political control of its Arctic territory. The Danish parliament – the *Folketing* – ratified the 1941 defence agreement in 1945 as its first post-war foreign policy act. It did so because it expected the agreement to be terminated (Petersen 2007, 2011). US strategic planners and senior military officers had other ideas. General H.H. Arnold is famously quoted as saying that if a third world war were to occur, "its strategic center [would] be the North Pole" (Grant 2010: 286). As Farish (2010: 175) has put it, "the north was understood as vital to continental defence against a gathering external threat." The shortest route for a possible Soviet attack on the North American mainland – and even a Soviet invasion – lay over the North Pole, and so Arctic regions such as northern Canada and Greenland were crucial zones for the construction of high latitude military surveillance and transport infrastructure necessary for both offensive operations and defensive abilities. The US recognised that Greenland and the wider Arctic would continue to matter to the security of the northern hemisphere.

In response to Danish efforts to terminate the 1941 agreement, the Americans invoked its Article X, which stated that:

> This agreement shall remain in force until it is agreed that the present dangers to the peace and security of the American continent have passed. At that time the modification or termination of the agreement will be the subject of consultation between the Government of the United States of America and the Government of Denmark. After due consultation has taken place, each party shall have the right to give the other party notice of its intention to terminate the agreement, and it is hereby agreed, that at the expiration of twelve months after such notice shall have been received by either party from the other this agreement shall cease to be in force.[4]

So, in the aftermath of World War II, US forces remained in Greenland because of its strategic importance and the lingering post-war dangers and emerging Cold War threats. Harry S. Truman's administration offered to buy Greenland in 1946 – a move which was rejected by Denmark – but also proposed a new agreement that would entrust the US with the territory's defence and allow the Americans to extend existing bases and build new military infrastructure (Olesen ibid.).

In 1951, a new defence agreement was signed by the Danes and the Americans. The North Atlantic Treaty Organization (NATO) had, in effect, precipitated this by requesting that the US and Denmark should make arrangements for military facilities in Greenland to be available for use by the armed forces of other NATO parties in defence of the area under the North Atlantic Treaty, which was signed in 1949. The 1951 agreement was, in a sense, an implementation of the treaty, and reassurances were given by the US that this would not infringe upon or threaten the sovereignty of the Kingdom of Denmark. As with the agreement of 1941, this new agreement allowed the US to construct, install, maintain, and operate facilities and equipment, including meteorological and communications installations and to store supplies; to station and house personnel and to provide for their health, recreation, and welfare; to provide for the protection and internal security of the area; to establish and maintain postal facilities and commissary stores; to control take-offs and landings, anchorages, moorings, and the movements and operation of ships, aircraft, and water-borne craft and vehicles, with due respect for the responsibilities of the Government of the Kingdom of Denmark in regard to shipping and aviation; and to improve and deepen harbours, channels, entrances, and anchorages. Both agreements allowed the US to enact processes of territorial enclosure.

One of the most significant outcomes of the agreement was the construction of Thule Air Base at Pituffik in the heart of the Inughuit homeland, which began in 1951. The base was operational by November 1952, was completed in 1953, and the final stage led to the dispossession of Inughuit. Some 116 people – making up around 27 families – were relocated in haste by the Danish state to Qaanaaq in 1953 because of the enlargement of the base into the area around the settlement of Uummannaq (also known as Dundas). Brøsted and Fægteborg (1987) discussed how the construction of the base infringed on Indigenous understandings of territory, affected the ecosystem on which Inughuit depended, and led to economic hardship. They also argued that it was inconsistent with the principles of the 1951 defence agreement, broke international law, and led to the unconstitutional confiscation of Inughuit property – an issue to which I return at the end of this chapter. The relocation, as Denmark saw it, was necessary for reasons of security; yet, for the Inughuit

Figure 4.1 Thule Air Base was constructed between 1951 and 1953 following the 1951 defence agreement signed by Denmark and the US. It was renamed Pituffik Space Base in April 2023 (Photograph Mark Nuttall).

who were forced to leave, it was an eviction. Already having experienced the impact of the construction of the base on their lives and surroundings, they were now removed from their homes and had to shelter in tents while they waited for Denmark to construct new houses for them.

Petersen (ibid.) argued that the establishment of Thule marked the beginning of a US polar strategy. During the early decades of the Cold War, northern Greenland became a strategic and geopolitical region for activities of surveillance and observation. The US had established a radio and weather station at Dundas in 1943 – it was known as Bluie West Six – and a gravel airstrip and balloon weather observatory were added after World War II. The station was eventually moved to Pituffik, when the air base was constructed. One of the original purposes for Thule Air Base was for it to be a staging base for the bombers of US Strategic Air Command (SAC). Originally, these were B-36 and later B-47 aircraft, which would be based in the US and use Thule when flying across the Arctic Ocean. But Thule was also important for reconnaissance aircraft which often penetrated Soviet airspace, including flights over the Franz Josef and Novaya Zemlya archipelagos. President Dwight D. Eisenhower had put forward a proposal for an Open Skies initiative in 1955, which was an idea for a system of aerial observation as part of a strategy to contain and ultimately defeat the Soviet Union in the event of war. In spring 1956, Thule was vital as a base for a large operation that resulted in the mapping by the US Air Force of the northern Soviet Arctic coast from the Kola Peninsula to the Bering Strait.

By the late 1950s, the B-52 bomber was in service. Its intercontinental range meant it did not need inflight refuelling or to use Thule as an intermediate base, while reconnaissance of the Soviet Arctic could now be done by the Lockheed U-2 high altitude aircraft, which were first deployed in 1956, and by satellites. SAC's units were withdrawn from Thule by mid-1959, but the air base remained critical for strategic defence. The Soviet Union had developed its own intercontinental bombers and the world's first intercontinental ballistic missile (ICBM), the R-7, which had flown more than 6,000 kilometres (3,700 miles) in a successful test in August 1957. The first R-7 unit became operational in February 1959 at Plesetsk in Mirny in northwest Russia. Thule became an essential early warning station for Soviet bomber attack missions and ICBM launches. The Ballistic Missile Early Warning System (BMEWS) was constructed in 1960–1961, with the main radar located at Thule and two smaller radars in Alaska and Britain. The construction of radar stations continued to be essential to North American defence strategy. Canada and the US built the Pinetree Line in 1946, and the Mid-Canada Line and the Distant Early Warning (DEW) Line in the 1950s.

Scientific investigations were critical to the success of these military operations. Geologist and glaciologist Richard Foster Flint wrote in one major report that:

> ...the arctic region is a part of the North American frontier that requires defense and that possesses strategic importance. The shortest great-circle distances between North America and the Old World cross it, involving air operations and requiring their logistical support. Any future war involving

the United States or Canada is almost certain to include operations in the Arctic. Although such operations may not constitute a major element in the total effort, nevertheless, from the strategic point of view they may well be of major importance.

(Flint 1953: 2)

Given this, Flint pointed to the need for increased understanding of polar environments. Canada and the US had initiated the Joint Arctic Weather Station (JAWS) programme in 1947, while Canada created the Advisory Committee on Northern Development (ACND) in 1949 to monitor Cold War-related events in the Arctic. Weather and radio stations were established, and while airfields and bases were essential to US Arctic defence strategy, they also served as hubs for scientific work, such as the Arctops (Arctic Topography) Project which gathered scientific data about Arctic terrain and resources. But as Flint (ibid.) observed:

...many problems could not be solved because of lack of knowledge of the basic properties of snow and ice. These substances had not been subjected to adequate physical analysis. When the runners of a ski-equipped airplane slide over snow with much less friction at one time or place than at another, it is essential to know why and to know how the difference can be controlled, because in many cases ability to take off is involved. To learn the reasons underlying the behavior of skis on snow, it is essential to understand the physical properties and phase relations of snow crystals, and this requires painstaking research.

Early efforts were put into an investigation of the design, construction, and maintenance of aerodromes on ice, which was conducted by the US Corps of Engineers' Soils Laboratory in Boston during 1946–1947. Research initiated there focused on the formation and melting of ice, the characteristics of snow, and the bearing capacity of ice under heavy aircraft. Suitable aerodrome sites in the Arctic and sub-Arctic needed to be identified and the topography mapped. In 1947, as part of this investigation, the Atlantic Division of the US Air Force's Air Transport Command initiated Project Snowman. Its objective was the development of appropriate survival and rescue techniques in Greenland, an assessment of the possibilities for aircraft operations, and an advancement of the engineering knowledge needed for the construction and maintenance of aerodromes on the inland ice. A number of preliminary studies were conducted and reconnaissance flights were made over the inland ice from Søndre Strømfjord.

A team comprising Air Force officers and personnel and observers from the Corps of Engineers landed on the inland ice in a ski-wheel C-47 aircraft and established a camp. They spent almost a month there, making observations and investigations and conducting tests. Detailed reports were published on the engineering aspects of the studies, the tests conducted, the results, and recommendations for future work. Project Snowman's investigations indicated that airfields for wheeled planes of virtually any weight were feasible in certain zones of the inland ice for an estimated nine months of the year. A supplementary report on "Aviation uses of ice" was written by Arctic anthropologist Vilhjalmur Stefansson and English translations of Russian reports on sea ice, air expeditions to the Arctic, and airfields on ice were appended to the main report.[5]

Scientific enquiry in northern Greenland was inspired by American military interests to ensure the safe and efficient movement of aircraft and shipping as well as understand the nature of ice for wartime operations. Greenland's inland ice was just one of many elements of the Arctic environment under scientific investigation and strategic audit. Environmental historian Janet Martin-Nielsen has written how, between 1948 and 1966, US forces in Greenland were caught up in what she refers to as the 'other cold war' – in this case, it was the inland ice that presented a quite daunting challenge, and efforts went into getting to know it and how to work with it to allow military action and surveillance (Martin-Nielsen 2012). The Americans realised they needed to co-operate with the ice rather than fight against it and make any attempt to subdue it. To be able to do this successfully, planners and strategists needed detailed knowledge and understanding of Greenland icescapes, landscapes, and seascapes. Resources and research effort, therefore, went into military programmes to understand what permafrost, sea ice, glacial ice, winds and prevailing weather systems meant for airstrips, road construction and maintenance, flights, ship and submarine transit, and navigation and troop movement (Dodds and Nuttall ibid.).

The establishment of the US Army's Snow, Ice and Permafrost Research Establishment (SIPRE) in 1949 was key to this work. SIPRE moved to Wilmette, Illinois, in 1951. Max Weber characterised bureaucracy as having fixed jurisdictional areas that are ordered by rules and regulations, but he recognised that bureaucracy also requires the performance of regular activities so that it can reproduce itself and extend its scope. SIPRE was critical to the way US military bureaucracy sought to understand Greenland's icy environment. Its reports were vital technological elements in support of the administrative apparatus responsible for the Thule Air Base area and they were essential documents in support of US military aims, but SIPRE initiated research projects and activities that saw American military presence in northern Greenland stretch out beyond the immediate boundaries of the defence area the base circumscribed into the empty spaces of the ice sheet, the coastal waters, and up into the skies.

In 1953, SIPRE was invited to participate in Project Mint Julep, which was organised by the American Geographical Society under a US Air Force contract. The aim was to assess the feasibility of using the southern area of the Greenland ice sheet to construct landing strips for military aircraft (Schuster 1954). In Northwest Greenland, SIPRE's Operation Icecap programme began to investigate the surface of the inland ice as well as peer within it. Party Crystal, which was a small mobile scientific team, worked eastward from Thule Air Base from May to September 1954 measuring elevation, snow properties, snow accumulation, density, stratigraphy and permeability, summer melt, and drilling ice cores (Benson 1955). The data acquired by Party Crystal were put to use in considerations for transportation and engineering operations on the inland ice, including excavations, digging, and tunnelling. Much of the scientific work was also necessary for informing plans for the further expansion of military activities and the construction of other bases or access to emergency landing strips at Station Nord in Northeast Greenland, where a joint Danish-American weather station had been established in 1952 as part of the wider network of radio and weather stations in the High Arctic.

Between 1952 and 1954, a SIPRE project carried out research on the inland ice that focused on the structure of névé, which is granular snow that accumulates and eventually becomes compacted into glacial ice (Benson 1959). Under the leadership of the Swiss-born Henri Bader, SIPRE scientists worked closely with US Army personnel and undertook research on the nature and movement of snow and glacial ice, and the stability of the inland ice. Bader initiated scientific research on ice cores, which were transported from Greenland, using dry ice for refrigeration, to cold rooms in Wilmette. At SIPRE, scientists archived the cores, measured accumulation layers and studied their physical and chemical composition (Dodds and Nuttall ibid.). In South Greenland, a short field reconnaissance carried out in summer 1956 was supplemented by maps, aerial photographs, and geological reports to locate an area that could be suitable for an overland route to – and allow year-round access on – a section of the inland ice for military purposes from Narsarsuaq airbase as well as the possible construction of additional piers, anchorages, and storage facilities for heavy equipment (Frost 1957). Knowledge of mountains, geology, glaciers, and the topography of the inland ice – in particular, the location of thaw lakes, meltwater streams and crevasses, glacial debris, and fine-grained soils as well as summer slush conditions – were crucial to inform a decision about the selection of the road.

Further studies that were key to understanding barriers to the mobility and movement of military operations were concerned with such matters as fog whiteout on the inland ice (Reiquam and Diamond 1959). The US military was particularly interested in travel conditions in the ablation zone, which includes those edges of the ice sheet where summer melt exceeds winter snow accumulation. A report on the environment of the inland ice on military logistics and performance assessed crevasses, sastrugi, melt streams, and ice ridges, and considered them to be slick and extremely rough for vehicular traffic. Cryoconite, which is a deposit of dust, soot, and microbes, and found in small pockets on the inland ice, was also subject to scientific investigation. The report concluded that while cryoconite holes did not appreciably hinder the movement of tracked vehicles, they were treacherous for people on foot, especially if the holes were obscured by blowing snow (Hogue 1964).

Probing the subsurface

In summer 1953, a delegation from SIPRE visited Thule to consider the engineering and associated problems that needed to be solved if the US military exploitation of Greenland's geostrategic location were to be possible. By this time, the strategic value of the Greenland inland ice, which had been the subject of a number of reports, was established. The visit identified a number of research objectives for the development of concepts, methods, techniques, and equipment essential for the construction of camps on ice caps and the inland ice. Much of this was to focus on the use of snow as construction material, foundations in snow and ice, Arctic housing, the control of drifting snow, sources of power, water supply, and waste disposal. In 1954, a Corps of Engineers research and development programme was initiated to address these problems.

Specific research and development tasks were assigned to laboratories under the Corps of Engineers, including SIPRE, the US Army Engineer Research and Development Laboratories (ERDL), the US Army Engineer Waterways Experiment Station (EWES), and the US Army Engineer Construction and Frost Effects Laboratory (ACFEL). SIPRE and ACFEL were merged to form the US Army Cold Regions Research and Development Laboratory (CRREL) in February 1961. The first Engineer Arctic Task Force (EATF) was assigned to support the Greenland activities, which included the construction and operation of camps and facilities and the transportation of personnel and equipment that the laboratories needed (Clark 1965). EATF was redesignated the US Army EATF in 1956, became the US Army Polar Research and Development Center (PR&DC) in 1958, and was later redesignated the US Army Research Support Group.

Central to the research activities concerned with operations on ice was the question of whether to construct ice cap camps on the surface or beneath it. Until the late 1950s, all installations on the inland ice had been constructed on the surface. At the time, the movement of ice was not thought to be a consideration, or at least something that was challenging and troubling to structures, but storms, drifting snow, and low temperatures were thought to be serious disadvantages to surface facilities. The accumulation and continuing drifting of snow meant that any structure built on the surface soon became buried, with the overburden placing it under great stress. One solution mooted was to place installations on jacking systems so that they could be raised above the surface of accumulating snow. Apart from the cost of construction and maintenance, this was dismissed as impractical as it would expose military buildings and other installations, making them highly visible and difficult to camouflage.

The subsurface camp concept was considered to be the best solution. Underground structures could be built, it was thought, to withstand the overburden of snow. They would be easier to camouflage and less vulnerable to enemy attack. Yet, this presented significant challenges because it required tunnelling into snow and ice and erecting buildings within the tunnels. And little was known about how the overburden of snow would even affect a subsurface structure. Only a few small tunnels driven for short distances into glaciers had been previously made, and then only for the purpose of glacial studies, not for permanent military use (Rausch 1958). Theories were put forward that there would be an optimum zone in which density, temperature, and deformation rates experienced close to the surface and even at greater depths would be lower, thus minimising the damage to subsurface buildings.

The military utilisation of the Greenland inland ice required year-round access to camps and other facilities. The development of skis for cargo planes and improved snow tractor technology meant air and surface transport increased the reliability of access in the summer. However, darkness, cold temperatures, and violent storms made such transport hazardous during winter. The US considered subsurface roads that would give year-round access to camps anywhere on the inland ice. However, a subsurface road network would only be feasible economically if a number of major installations were located on the ice. SIPRE turned its attention to researching the possibilities of constructing tunnels and rooms under the ice for a variety of purposes, including the storage of military equipment, armaments, and fuel. This resulted in a series of projects and reports on the excavation, development, and maintenance of military facilities in permanently frozen ground.

These activities needed a land route approach to the inland ice and so in 1954 the US Army Corps of Engineers constructed Camp TUTO (Thule Take-Off) at the edge of the ice, some 18 miles from Thule Air Base (in this chapter, I also use imperial rather than metric measurements, as it was the imperial system used in the operations and in the reports). A gravel road from Thule and a ramp up on to the ice capable of supporting heavy military traffic gave year-round access, although it was found that the life of gravel roads was limited by the significant amount of ablation of the surface of the inland ice adjacent to them as well as by meltwater flow that affected erosion. This presented a challenge to maintaining roads on the ice by minimising dust settlement and by diverting melt water. Other access to the ablation zone was provided by Nuna Ramp, 26 miles northeast of Thule, and research focused on the methods and techniques needed to build and maintain a larger network of roads on boulder and permafrost terrain (Linell et al. 1956). Also in summer 1954, a team of engineers carried out a field reconnaissance to determine the feasibility of constructing a high pressure petroleum products pipeline from the Thule area (utilising either Camp TUTO or the Nuna Ramp) to a number of points on the ice sheet. Although a pipeline was never built, the report concluded that there were no known insurmountable problems in constructing, operating, and maintaining one, along with the associated infrastructure needed for fuel storage, transport, pump station installations, and living facilities for the operating crews (Karstens, Nelson and Kearney 1954). That same year, trenching capabilities on the ice were tested and lightweight portable structures were installed in the excavations and tunnels (Waterhouse 1955), while experimental tunnelling was carried out during the summers of 1955 and 1956 near Camp TUTO (Rausch ibid.). During the first summer's work, a tunnel was driven for 500 feet and a 34 feet room was excavated – the tunnel was extended by a further 125 feet the following year and the room was enlarged.

The first deliberately designed subsurface camp was constructed in 1957 by a US Army Engineer task force at Camp Fistclench, 220 miles east of Thule. All this required the construction of access roads on the ice, while storage facilities were assembled at the site for generators, compressors, and explosives. Mining engineers were deployed to carry out the work, and they used conventional mining methods to blast and drill their way into the ice and drive tunnels in which steel arches and roof supports were installed. John Abel Jr, for example, authored a report about an experimental tunnel that was driven into a glacial till hillside near Camp TUTO during summer 1959. Nine feet wide and ten feet high, it extended for 300 feet. It was an experiment to determine the feasibility of excavating subsurface openings in ice and frozen ground (Abel 1960). In carrying out such work, Abel and many others like him acted in the role of the engineer-*bricoleur*, engaging with "uneven, unruly, and unstable environments out of which infrastructures are made" (Harvey and Knox 2015: 7).

Understanding these military activities and scientific endeavours in northern Greenland during the Cold War requires engagement with a vast archive of documents produced by SIPRE and the other agencies involved, and which is made up of scientific reports and engineering feasibility studies. This archive contains facts and knowledge about northern Greenland's inland ice and its coastal waters. We can also read these documents and reports in another way, as they present data that form something of an important baseline for our understanding of glacial ice and ice sheets as well as a changing climate and its effects on the Arctic. These documents had a bureaucratic purpose, too, being essential for the construction and enclosure of northern Greenland as a US-administered region that stretched beyond the boundaries of the immediate Thule Air Base defence area into the empty spaces of the ice sheet, to Camp Century, Camp Fistclench, and other areas, out into Baffin Bay and up into the skies.

In the 1950s, US Army units undertook a range of initiatives to explore the possibilities of establishing structures inside the ice as places for safety and refuge or to locate missiles. Convoys of tracked vehicles that formed supply trains criss-crossed the inland ice. This mode of transportation was used to take the equipment needed from Thule to construct the weather station at Station Nord on the northeast coast, a joint venture between Denmark and the US. Some of the scientists and engineers working at SIPRE and later CRREL were involved in the construction of Camp Century, some 150 miles northeast of Thule. This drew upon the intensive research on the ice sheet that had been carried out during the 1950s and cut-and-cover trenching techniques were used to build a scientific military subsurface base that could accommodate up to 200 military personnel. It was powered by a nuclear generator and a subsurface railway was also installed. As Nielsen, Nielsen, and Martin-Nielsen (2014: 445) put it:

> Camp Century was part of the US Army's new concept for polar military engineering. Transforming the snowy wasteland into a veritable city equipped with every convenience from library to warm showers, the Army saw Camp Century as a stepping-stone to increased military presence on (and, quite literally, in) the icecap.

The US Army Corps of Engineers used Camp Century to study and assess the feasibility of Project Iceworm, which would have involved the deployment of 600 ballistic missiles under the ice (Nielsen and Nielsen 2021). The Danish government was never informed at the time about the real purpose of Project Iceworm – the official story about Camp Century was that it was place where American engineers were experimenting with cold weather engineering techniques and where scientists sought to understand the nature of ice. Indeed, this is precisely what was going on there (Clark ibid.), and the project was "also described as a techno-scientific conquest of the Arctic, which ultimately could be seen as a first step to outer space" (Nielsen, Nielsen and Martin-Nielsen ibid.: 461), but it did provide a convenient cover story for explaining other military studies. While Project Iceworm remained secret and all documents about it were classified, there was public information disseminated about Camp Century in the US, including films about its construction and about life in what was called a 'city under the ice.' Project Iceworm was never realised and year-round activities at Camp Century ended in 1964, with seasonal work there coming to a halt in 1967, when the facility was abandoned – the nuclear generator having first been removed (Colgan et al. 2016; Nielsen and Nielsen ibid.; Weiss 2011).

Camp Century had originally been given a life term of around ten years because of the calculations about ice movement. Greenland was not thought to be inert, and so it was not a surprise that the ice would eventually begin to affect the city beneath the ice. Finnish-Swedish geologist, mineralogist, and polar explorer Alfred Erik Nordenskiöld had first visited the inland ice in 1870, and in his account of the expedition, he wrote how it was in constant motion, advancing slowly, but with a different velocity in different areas. He also reported how the movement in the ice gave rise to huge chasms and clefts. And as James Geikie wrote in 1874, "when we think about the immense extent of the glacier system of Greenland, and how in the interior every hill is covered and every valley filled to overflowing with a moving sea of snow and ice, we can hardly overestimate the tremendous tear and wear to which the buried country must be subjected" (Geikie 1874: 58). The Army's scientists factored in what they knew about this movement when Camp Century was being planned and designed, but were surprised by how quickly the ice closed in and started to crush the roof and walls only after a couple of years following its construction.

Figure 4.2 Scientific research on Greenland's inland ice, glaciers, sea ice, and geology was critical for the success of military operations during the Cold War (Photograph Mark Nuttall).

Observing and controlling northern seascapes

From the 1950s onwards, Greenland's waters from the west coast to the north around Thule and beyond were considered to be vital sea lanes for naval, coast guard, and merchant vessels responsible for establishing and supplying the infrastructure and installations of meteorology, radar systems, and defence in the High Arctic. However, ice often hindered, obstructed, and frustrated the smooth flow of maritime traffic. Oceanographic data from Baffin Bay, Melville Bay, and the seas of the Canadian Arctic Archipelago and northern Greenland had been acquired on various discovery expeditions in the late nineteenth century as well as specific scientific voyages during the early twentieth century (for a survey, see Muench 1971). However, American military activities during the immediate post-World War II period and the early decades of the Cold War demanded a more complete understanding of the properties of ocean, sea ice, and tidewater glaciers for the establishment of marine routes. Considerable research effort and the investment of resources went into tracking, sizing up, and measuring ice and icebergs, studying ocean depths, and surveying, mapping, and charting the possibilities of transiting safely and efficiently through more fluid, ice-free passages.

Along with scientific activities that were associated largely with military or security operations, Melville Bay and areas further north (such as the North Water polynya) were also sites of interest for oceanographic research to understand water flow, heat budgets, and ice formation. During the last couple of decades of the Cold War period, geological surveys of the seabed aimed for greater understanding of its extent and nature, but they also revealed something about the region's hydrocarbon potential, and that data established a baseline for later resource exploration. Large-scale surveillance programmes and research initiatives ensured Northwest Greenland was a busy region – military operations, aerial reconnaissance flights, and scientific expeditions meant the skies and seas were criss-crossed extensively – yet, as I discuss later, the Indigenous inhabitants were not necessarily informed about such activity; nor were they consulted about the nature of ice and water, clouds, winds, and weather patterns.

As with the geophysical sciences, oceanography became essential to Cold War scientific endeavours to understand the Earth (Dennis 2003; Turchetti and Roberts 2014), and so attention was also turned to the surveillance, observation, and study of sea ice (Roberts 2014) and to understanding the depths of Arctic seas. From the 1950s, research on the physical oceanography of the wider north Baffin Bay area, the North Water polynya, the waters stretching to the Nares Strait region and beyond, and the Arctic ice zone, became a priority for the US Navy. Data were needed for safer navigation, the transit of icebreakers and supply ships, and the smooth flow of personnel, supplies, and equipment to and from northern Greenland and the High Arctic. But scientific research was also important for gaining a greater understanding of coastal geographies, ice fields, the volumetric space and depths of northern waters, the shape and formation of the seabed, and the potential for the discovery of resources. Dennis points out that field sciences, such as oceanography, "take place in spaces that are neither easily contained or controlled. Nonetheless, the need to contain and control those spaces drives much of the research" (Dennis ibid.: 809). For the US military, it was vital to be able to see what lay ahead in the sea, on the surface in terms of charting the location and extent of ice and icebergs, and over towards the horizon, but it was just as critical to know what was within the sea, such as the submerged parts of icebergs.

In *War and Cinema*, Paul Virilio (1989) argued that war is less about victories on battlefields and securing territories and more about how innovations in technology, particularly how technologies of representation could influence and shape the visual perceptions of the battlefield and military action. For Virilio, advances in technology, whether through radar images, maps, and other forms of representation, were all key in military and scientific efforts to make a 'dense country' knowable and 'transparent.' As Doel, Levin, and Marker (2006) describe, when Bruce Heezen and Marie Tharp from Columbia University created the first comprehensive physiographic map of the North Atlantic Ocean basin in 1957, they drew on depth profiles of the seafloor and other information derived from US military-funded oceanographic research carried out during the early years of the Cold War. Mapping and representing the seabed and oceanic troughs and ridges reflected particular kinds of motivations for doing such work in the first place.

During the Cold War, North Greenlandic waters and the Arctic ice zone, including Melville Bay, came to be viewed, like other northern ocean spaces such as the North Atlantic, as "a vast topography for military surveillance" (cf. Mac-Donald 2006). They also became spaces for military transit to supply and secure operations at Thule Air Base. The 'dense country' of pack ice, icebergs, glaciers, fjords, inlets and islands, and the murky glacial waters themselves needed to become transparent. As essential areas for maritime patrol and transport systems, Baffin Bay and Melville Bay became Cold War geopolitical seascapes (cf. Mac-Donald ibid.) during an era when oceans were becoming subject to initiatives to turn them from wild, unruly, vast stretches of water to knowable and controllable spaces.

From the end of the Second World War, the Soviet navy had developed from a coastal defence force to an ocean-going one. NATO's response was to monitor Soviet naval vessel activity at what were known strategically as choke points, or narrow waterways. This was relatively easy to do in the Baltic Sea, where Soviet vessels had to pass through NATO-controlled straits such as the Kattegat and Skagerrak into the Atlantic, or in the Black Sea region through the Bosporus and Dardenelles into the Mediterranean. However, there was no such natural choke point governing the exit of ships and submarines from the Northern Fleet's home ports on the Kola Peninsula into the North Atlantic. The northern North Atlantic, especially the area from the Norwegian Sea to the waters between Greenland, Iceland and the United Kingdom became a new defensive line (Critchley 1984). On the east coast of Greenland, for instance, and into the North Atlantic, the Greenland-Iceland-United Kingdom (GIUK) Gap was a choke point that had been vital for Allied naval and merchant fleets during World War II. It was critical for allowing the flow of military and commercial shipping, but also for restricting navigation by German fleets. The strategic importance of the GIUK Gap continued into the Cold War period. It was a crucial area for ocean surveillance, especially for monitoring Soviet vessels and submarines from the 1950s through American and British efforts (Robinson 2014), and it remains significant in this contemporary age of geopolitical tension and climate change.

In particular, during the Cold War strong anti-submarine defence measures (including the installation of underwater sensor and listening posts) were taken by the NATO allies to detect, deter, intercept, and prevent movement of the Soviet Union's nuclear ballistic missile submarine force (Alexander and Morgan 1988). The assumption was that the fleet had to pass through the GIUK Gap to be able to launch missiles at the eastern seaboard of the US, and some contemporary scholars and military strategists continue to emphasise the importance of controlling the GIUK Gap today to secure UK and US defence systems. As Østreng (1977) pointed out over forty five years ago, however, advocates of the GIUK Gap monitoring and detection system overlooked the technological capabilities of Soviet nuclear ballistic missiles at the time to cross wider distances and he suggested that the Arctic Ocean may have served the Soviets just as well as a missile launching and transit area.

Surveys of Baffin Bay and the waters and ice further north extended this oceanographic knowledge and panoptic reach. Oceanographic knowledge of the depths and bathymetry of Melville Bay and understanding of the shapes and contours of the sea bed, but also knowledge of its glaciers and mountain heights and how the inland ice affected cloud formation and winds was necessary to make it legible and less dangerous and problematic when producing weather and ice forecasts and planning routes and plotting a course through and across it by ship as well as above and over it by aircraft. By making Baffin Bay and Melville Bay transparent, it was important to be able to see across the water, but also within, and so making it transparent required a perspective from which it became possible to model the sea to visualise its volume, depths, and appearances. Gísli Pálsson's idea of how the sea is presented as a virtual aquarium (important, he argues, for how fisheries managers are able to visualise and measure fish stocks and deep sea ecosystems; see Pálsson 1998) seems apt for considering how northern Greenland's waters were modelled and how perceptions of them were shaped by technologies and practices of observation, surveillance, the measurement of water columns, and the mapping of the sea bed. Being able to look within the darkest depths of the sea was only part of the process of knowing the Arctic, however; ice observers looked down on ice, sea, glaciers, and ice sheets, and extended their vision through the skies, air, and clouds, and up into the atmosphere.

Securitisation, Denmark as an Arctic state, and Indigenous communities

The construction of air bases as well as scientific research carried out on the inland ice and on the nature and properties of sea ice and icebergs as well the geological mapping of Arctic terrain meant that much of northern Greenland became a securitised region that was effectively administered by the US. Parts of the country, most notably the Thule Air Base defence area and large stretches of the ice sheet, were marked off as militarised landscapes in which experimental work was done on ice and snow to assess the possibilities of tunnel and camp construction, road building, the operation of bomber aircraft and military vehicles in polar environments, and the placement of missiles deep in the ice. While these were considered sites of possible combat, they were largely sites of war preparation rather than ever becoming icy battlefields (cf. Coates et al. 2011).

The defence agreements of 1941 and 1951 show how Greenland has been central to Danish foreign strategy since World War II. Greenland – and the Faroe Islands and Iceland – were of geostrategic importance to the Allied nations and for control over the northern North Atlantic, and they remained vital for northern hemispheric security during the Cold War. The naval and air base at Keflavik in Iceland, for example, was a key node for securitisation efforts. Initiatives intended to test and maintain missile and bomb sites, monitor Soviet military movements, control airspace, and secure sea routes led to the building of Thule Air Base and infrastructure that often entailed ecological rupture and forms of cultural violence, dispossession, and displacement. Enlargement of the defence area – to include a new weather station at the site of the community's settlement, the previous station having been moved earlier to Pituffik when construction on the base started (Brøsted and Fægteborg ibid.) – meant eviction of Inughuit from the community of Uummannaq and the erasure of Indigenous human presence there in order to ensure American security. The area became renowned as a place of dispossession. To say this was deeply unsettling does not do justice to how they – and other Indigenous and local communities around the world – experienced the power and reach of the state through narratives of fear, the creation of scientific-technological landscapes, and expulsion.

Despite projects such as Camp Century being presented to the public as the scientific conquest of the Earth's remote, icy, and empty places, Greenland's Cold War environments were also marked out and prepared as environments of violence, in the sense of them being potential battlefields, or from where violence would be enacted in the event of conflict and war (in the case of missile launch sites), or they were caught up in and affected by accidents and contamination (such as the crash of a US Air Force B-52 bomber that was carrying four nuclear weapons on the sea ice 12 kilometres from Thule in January 1968). Thule Air Base is now known as Pituffik Space Base and it continues to be vital to US military strategy and it influences Danish-American relations. For the Inughuit of the region, however, the base stands as a reminder of forced movement, environmental ruin, and ecological crisis (Hastrup 2019).

There are many residents of Qaanaaq and other villages in Avanersuaq who remember the relocation from Uummannaq and the trauma it entailed. There are many who also worry about the toxic legacies of Camp Century and the B-52 crash. I have sat in people's homes and heard them talk about the musk ox and the seals they have caught that have deformities and cancerous tumours they attribute to the activities of the American military during the Cold War. What else is there, they ask, that the Americans did on the great ice, on *sermersuaq*, that we don't know about? What is there that even Denmark does not know about?

In spring 2015, I sat in the home of one hunter, who was also something of an activist for the rights of Inughuit, especially those who were moved to Qaanaaq in 1953, and listened as he told me of his suspicions that he was being tracked by the US military. He was convinced his telephone landline was tapped, as he heard a mysterious click each time he used it. Anxieties over toxic legacies are heightened now that Camp Century and other ice sheet bases are beginning to emerge from the ice in a warming climate, leading to the eventual remobilisation of abandoned physical, chemical, biological, and radiological waste (Colgan et al. ibid.). And determining who is responsible for the remediation and clean-up at these sites is a matter of political dispute.

As Greenland has been – indeed, continues to be – vital to the air defence of North America and Europe, and as both the US and Denmark are NATO members, much remains unknown about the history of Thule Air Base and its operational mandate. The defence area remains critical territory for the US and for geo-security, but it also remains crucial for Danish-American relations. As an Arctic state, it is through Greenland that the Kingdom of Denmark is able to position itself as a key player in world affairs. Danish relations with the US must be understood in part through historical and current US geopolitical interests in Greenland and how far Denmark is able to influence America's position on the Arctic. But Greenland's rapidly evolving political landscape, its economic development strategy, and its aspirations for independence play a critical role in Danish-Greenlandic relations as well as for the forging of bilateral relations between Greenland and the US. In October 2020, the US and Greenland signed an agreement on developing stronger economic relations as well as security. It covers bilateral trade and investment, science, minerals and energy, and economic growth through areas such as tourism and education. What is significant to note here is that the agreement is termed a common plan within the framework of existing co-operative relatives regarding Pituffik Space Base.

Greenland has no military of its own. From its base in Nuuk, the Danish military presence of Arctic Command continues to be responsible for fisheries protection and search and rescue (it acts as the maritime rescue co-ordination centre for the Greenlandic search and rescue region) as well as with surveillance, defence, and maintenance of territorial sovereignty, but has a broad mandate to carry out quasi-civilian tasks in connection with maritime activities. These include anti-pollution and spill recovery activities in the open ocean, providing ice-breaking support to local companies, carrying out hydrographic surveys, and monitoring commercial activities in Greenlandic waters. Related to these responsibilities and tasks, Arctic Command faces emerging challenges from increasing mining and resource exploration and development, new shipping routes, an increase in tourism (specifically in the form of greater numbers of cruise ships), and from an expansion of scientific research. Arctic Command was established in 2012 with a merger of Greenland Command and Faroes Command, which were essentially charged with coast guard duties. Greenland Command's base had been at Grønnedal; the Danish Navy having moved there from Godthåb when the US Navy returned the facility to Denmark in 1951. The Sirius Patrol, which originated as the dog sledge patrol established by the US Coast Guard in 1941, is also under the administration of Arctic Command. As a member of NATO, Denmark has also been involved in the formation of a Nordic alliance that includes fellow NATO member Norway with partners Finland and Sweden, and Greenland, the Faroe Islands, and the Åland Islands (an autonomous region of Finland) for joint monitoring of Nordic marine areas, Nordic air space, and the Arctic.[6] Rahbek-Clemmensen (2011: 14) argues that Denmark's "central policy-makers have already realized that their own military means will never be enough to hold on to Greenland," and the importance of such regional institutions and alliances may be critical if Denmark is to maintain a presence in Greenland and retain its status as an Arctic nation.

In September 2020, NATO Secretary General Jens Stoltenberg drew attention to a paradox in the Arctic. He said that while it was a region characterised by several decades of co-operation – most notably through the work of the Arctic Council – "under the ice in the Arctic, you have some of the most dangerous weapons in the world."[7] Stoltenberg expressed his concerns over Russia's deployment of new missile systems in its Arctic regions as well as its nuclear submarine capabilities. He also spoke of the emergence of China as a naval power and the country's investment in icebreakers. China has defined itself as a 'near-Arctic' country and sees a 'polar silk road' as key to its Belt and Road initiative. China is an important market for Greenlandic fish products and has expressed interest in funding rare earth mining projects in Greenland. As the global rivalry of the larger powers intensifies, and as Arctic co-operation has weakened since Russia's invasion of Ukraine in February 2022, pausing the work of the Arctic Council, Greenland is placed once more at the centre of discussion about the Arctic as a strategic region, and both the subterranean and subsurface resources are entangled in new forms of geo-security. And while Stoltenberg highlighted missiles under the ice, it is the aerial dimension that becomes central to securitisation and territorial integrity (Williams 2010). Greenland is still Danish territory – at least, in how it constitutes part of the kingdom – and Denmark still has to assert control and power over its Arctic stretches.

In February 2021, the *Folketing*, the Danish Parliament, agreed on major funding for an Arctic capacity-building package that will see increased surveillance in Greenland through the deployment of drones and satellites, and with some of the money being set aside for defence education in the country. Ian Shaw writes that while capitalist systems of enclosure have privatised and secured spaces through means that territorialise power relations into the very soil, he argues that today enclosure operates through atmospheric spatiality. Aircraft, drones, satellites, and other technologies of airpower have enabled states to have control over the skies – what he calls vertical regimes of state power. Drones, for instance, produce new regimes of enclosure that fuse "militaristic forms of aerial occupation with the vertical logics of capital accumulation" (Shaw 2017: 884). So, while Denmark and Greenland are engaged in a process of mapping and surveying Greenland's coastal areas and its continental shelf, seeking to determine how far Greenland stretches out into the seabed structures of the Arctic Ocean and the North Atlantic, the atmosphere also becomes essential to understanding Greenland's territorial extent in this volumetric sense.

Indigenous people are noticeably absent from Cold War reports on ice, water, and the wider geographies of northern Greenland. Ice reconnaissance and shipping took place in an Arctic space that appeared empty of people. Knowledge of sea ice and weather was based initially on aerial, ship, and land-based observations, and later on, these methods of surveillance were combined with new meteorological techniques and practices, including radar and remote sensing. Surveys and observations, though, did not take note of Indigenous knowledge of changes in drift ice and fast ice extent, freeze-up and break-up patterns, glacier fronts and icebergs, and calving processes from glaciers, and how this facilitates both the formation and stability of the fast ice cover.

In my work in northern Greenland over the years, I have been struck by how precise and specific Indigenous knowledge of sea ice and icebergs is for travel and navigation – for example, knowing when one must go windward of an iceberg, when it is safe to go between icebergs, or how to identify a wave that is rising because a glacier has calved an iceberg, or because an iceberg has shifted its centre of gravity, or how to recognise whether a sudden gust of wind has come from a mountain or from along a fjord. Nor did later scientific interest in the 1960s and in the 1970s in the North Water and in the geological history of the Nares Strait region between Greenland and Canada (e.g. Dunbar 1969) take local understandings into consideration, despite there being a rich Indigenous vocabulary that hints at extensive knowledge of the movement of ice streams from the Greenland inland ice or that describes how icebergs have created troughs, ridges, passages, and furrows in the seafloor. By and large, for the people who lived along the coasts of Baffin Bay, within the boundaries of Melville Bay, and around the Avanersuaq district, or who travelled in those places to hunt and fish, the effects of Cold War activities were somewhat surreptitious. The effects and legacies, however, are apparent in the way people talk today about animals and places being agitated, and this is something I will return to in Chapters 6 and 7 when I discuss more contemporary probing and exploration of Greenland's subsurface and its waters. Before that, in Chapter 5, I consider how the representation and reproduction of remoteness matter in the making of spaces of extraction.

Notes

1 "President Trump eyes a new real estate purchase: Greenland." *The Wall Street Journal* 16 August 2019. https://www.wsj.com/articles/trump-eyes-a-new-real-estate-purchase-greenland-11565904223.

2 "Kuupik Kleist: The Cold War is re-introduced in Greenland." *High North News* 21 October 2019, https://www.highnorthnews.com/en/kuupik-kleist-cold-war-re-introduced-greenland. Accessed 26 January 2023.

3 This quote and those that follow come from the transcript of the press meeting with the delegation and is available on the US Department of State website at: https://www.state.gov/secretary-antony-j-blinken-greenlandic-premier-mute-egede-greenlandic-foreign-minister-pele-broberg-and-danish-foreign-minister-jeppe-kofod-at-a-joint-press-availability/.

4 Denmark-United States: Agreement Relating to the Defense of Greenland. Source: *The American Journal of International Law*, July 1941, Vol. 35, No. 3, Supplement: Official Documents (July 1941), pp. 132.

5 Published in May 1947, Vilhjamur Stefansson's report was called *The Aviation Uses of Ice* and was revised in May 1948.
6 "NATO's Arctic Military Alliance," http://www.globalresearch.ca/nato-s-arctic-military-alliance.
7 "Stoltenberg: 'Under the ice in the Arctic, you have some of the most dangerous weapons in the world.'" *High North News* 9 September 2020, https://www.highnorthnews.com/en/stoltenberg-under-ice-arctic-you-have-some-most-dangerous-weapons-world. Accessed 26 January 2023.

References

Abel, John F. Jr. 1960. *Permafrost Tunnel, Camp TUTO, Greenland.* SIPRE Technical Report No. 73, Wilmette, Illinois: U.S. Army Snow Ice and Permafrost Research Establishment.

Alexander, Lewis M. and Joseph R. Morgan. 1988. "Choke points of the world ocean: A geographic and military assessment." *Ocean Yearbook* 7 (1): 340–355.

Benson, Carl S. 1955. *Scientific Work of Party Crystal, 1954: Preliminary Report.* SIPRE Research Report 1954. Wilmette, Illinois: U.S. Army Snow Ice and Permafrost Research Establishment.

Benson, Carl S. 1959. *Physical Investigations on the Snow and Firn of Northwest Greenland during 1952, 1953 and 1954.* SIPRE Research Report 26. Wilmette, Illinois: U.S. Army Snow Ice and Permafrost Research Establishment.

Brøsted, Jens and Mads Fægteborg. 1987. *Thule – fangerfolk og militæranlæg.* Copenhagen: Akademisk Forlag.

Christiansen, Flemming G. 2022. "Greenland mineral exploration history." *Mineral Economics* https://doi.org/10.1007/s13563-022-00350-2.

Clark, Elmer F. 1965. *Camp Century: Evolution of Concept and History of Design, Construction and Performance.* Technical Report No. 174, Hanover, NH: U.S. Army Research and Engineering Laboratory.

Coates, Peter, Tim Cole, Marianna Dudley and Chris Pearson. 2011. "Defending nation, defending nature? Militarized landscapes and military environmentalism in Britain, France, and the United States." *Environmental History* 16 (3): 456–491.

Colgan, William, Horst Machguth, Mike McFerrin, Jeff D. Colgan, Dirk van As and Joseph A. Macgregor. 2016. "The abandoned ice sheet base at Camp Century, Greenland, in a warming climate." *Geophysical Research Letters* 43: 8091–8096, doi:10.1002/2016GL069688.

Critchley, W. Harriet. 1894. "Polar deployment of Soviet submarines." *International Journal* 39 (4): 828–868.

Dennis, Michael Aaron 2003. "Earthly matters: On the Cold War and the earth sciences." *Social Studies of Science* 33 (5): 809–819.

Dodds, Klaus and Mark Nuttall. 2016. *The Scramble for the Poles: The geopolitics of the Arctic and Antarctic.* Cambridge: Polity.

Dodds, Klaus and Mark Nuttall. 2018. "Materialising Greenland within a critical Arctic geopolitics." In Kristian Søby Kristensen and Jon Rahbek-Clemmensen (eds.) *Greenland and the International Relations of a Changing Arctic: Postcolonial paradiplomacy between High and Low Politics.* London and New York: Routledge, pp. 139–154.

Doel, R.E., T.J. Levin and M.K. Marker. 2006. "Extending modern cartography to the ocean depths: Military patronage, Cold War priorities, and the Heezen-Tharp mapping project, 1952–59." *Journal of Historical Geography* 32 (3): 605–626.

Dunbar, Moira. 1969. "The geographical position of the North Water." *Arctic* 22 (4): 438–441.

Farish, Matthew. 2010. *The Contours of America's Cold War.* Minneapolis: University of Minneapolis Press.

Flint, Richard Foster. 1953. *Snow, Ice and Permafrost in Military Operations.* SIPRE Report 15, U.S. Army Snow, Ice, and Permafrost Research Establishment. Engineer Research and Development Center. http://hdl.handle.net/11681/6000

Fogelson, Nancy. 1985. "The tip of the iceberg: The United States and international rivalry for the Arctic, 1900–25." *Diplomatic History* 9 (2): 131–148.

Franks, Jill. 2006. *Islands and the Modernists: The allure of isolation in art, literature and science.* Jefferson, NC and London: McFarland & Company.

Frost, Robert E. 1957. *A Reconnaissance for a Southern Greenland Ice-Cap Access for Military Purposes.* SIPRE Technical Report 46: Wilmette, Illinois: U.S. Army Snow Ice and Permafrost Research Establishment.

Geikie, James. 1874. *The Great Ice Age and Its Relation to the Antiquity of Man.* London: W. Isbister.

Grant, Shelagh. 2010. *Polar Imperative.* Vancouver: Douglas and McIntyre.

Hansen, Wallace. 1994. *Greenland's Icy Fury.* College Station: Texas A&M University Press.

Harvey, Penny and Hannah Knox. 2015. *Roads: An anthropology of infrastructure and expertise.* Ithaca: Cornell University Press.

Hastrup, Kirsten. 2019. "A community on the brink of extinction? Ecological crises and ruined landscapes in Northwest Greenland." In Astrid B. Stensrud and Thomas Hylland Eriksen (eds.) *Climate, Capitalism and Communities: An anthropology of environmental overheating.* London: Pluto Press, pp. 41–56.

Heidbrink, Ingo. 2022. "'No One Thinks of Greenland': US-Greenland relations and perceptions of Greenland in the US from the early modern period to the 20th century." *American Studies in Scandinavia* 54 (2): 8–34.

Heinrich, Jens. 2018. "Independence through international affairs: How foreign relations shaped Greenlandic identity before 1979." In Kristian Søby Kristensen and Jon Rahbek-Clemmensen (eds.) *Greenland and the International Relations of a Changing Arctic: Postcolonial paradiplomacy between High and Low Politics.* London and New York: Routledge, pp. 28–37.

Hogue, Donald W. 1964. *Environment of the Greenland Icecap.* U.S. Army Material Command, Technical Report E5–14. Natick, MA: U.S. Army Natick Laboratories.

Hough, Peter. 2013. *International Politics of the Arctic.* London and New York: Routledge.

Karsten, Chauncey W., Walter L. Nelson and Geo. W. Kearney. 1954. *Pipeline Study, Greenland.* Ft. Belvoir, Virginia: U.S. Army Engineer and Development Laboratories.

Latour, Bruno and Steve Woolgar. 1986. *Laboratory Life: The construction of scientific facts.* Princeton, NJ: Princeton University Press.

Linell, Kenneth, Charles W. Fulwider, Henry W. Stevens and A. Thomas Carroza. 1956. *Approach Roads Greenland 1954 Programme: Projects 1 and 10A.* Boston, MA: Arctic Construction and Frost Effects Laboratory.

MacDonald, Fraser. 2006. "The last outpost of Empire: Rockall and the Cold War." *Journal of Historical Geography* 32 (3): 627–647.

Martin-Nielsen, Janet. 2012. "The other cold war: The United States and Greenland's ice sheet environment, 1948–1966." *Journal of Historical Geography* 38 (1): 69–80.

Muench, Robin Davie. 1971. *The Physical Oceanography of the Northern Baffin Bay Region. The Baffin Bay-North Water Project.* Scientific Report No. 1. Calgary: Arctic Institute of North America.

Nielsen, Kristian. 2013. "Transforming Greenland: Imperial formations in the Cold War." *New Global Studies* 7 (2): 129–154.

Nielsen, Kristian H. and Henry Nielsen. 2021. *Camp Century: The untold story of America's secret army military base under the Greenland ice.* New York: Columbia University Press.

Nielsen, Kristian H., Henry Nielsen and Janet Martin-Nielsen. 2014. "City under the Ice: The closed world of Camp Century in Cold War culture." *Science as Culture* 23 (4): 443–464.

Olesen, Thorsten Borring. 2019. "Buying Greenland? Trump, Truman and the 'Pearl of the Mediterranean'." *Nordics Info* 10 September. https://nordics.info/show/artikel/buying-greenland-trump-truman-and-the-pearl-of-the-mediterranean.

Østreng, Willy. 1977. "The strategic balance and the Arctic Ocean: Soviet options." *Cooperation and Conflict* 12 (1): 41–62.

Pálsson, Gísli. 1998. "The virtual aquarium: Commodity fiction and cod fishing." *Ecological Economics* 24 (2–3): 275–288.

Petersen, Nikolaj. 2007. "Negotiating the 1951 Greenland Defense Agreement: Theoretical and empirical aspects." *Scandinavian Studies* 21 (1): 1–28.

Petersen, Nikolaj. 2011. "SAC at Thule: Greenland in U.S. polar strategy." *Journal of Cold War Studies* 13: 90–113.

Rahbek-Clemmensen, Jon. 2011. "Denmark in the Arctic." *Atlantisch Perspectief* 3: 9–14.

Rausch, Donald O. 1958. *Ice Tunnel, TUTO area, Greenland, 1956.* SIPRE Technical Report 44 U.S. Army Snow, Ice, and Permafrost Research Establishment. Engineer Research and Development Center. http://hdl.handle.net/11681/6010.

Reiquam, Howard and Marvin Diamond. 1959. *Investigations of Fog Whiteout.* SIPRE Research Report 52. Wilmette, Illinois: U.S. Army Snow Ice and Permafrost Research Establishment.

Roberts, Peder. 2014. "Scientists and sea ice under surveillance in the early Cold War." In Simone Turchetti and Peder Roberts (eds.) *The Surveillance Imperative: Geosciences during the Cold War and Beyond.* New York: Palgrave Macmillan, pp. 125–144.

Robinson, Michael F. 2006. *The Coldest Crucible: Arctic exploration and American culture.* Chicago: University of Chicago Press.

Robinson, Sam. 2014. "Stormy seas: Anglo-American negotiations on ocean surveillance." In Simone Turchetti and Peder Roberts (eds.) *The Surveillance Imperative: Geosciences during the Cold War and Beyond.* New York: Palgrave Macmillan, pp. 105–124.

Schuster, Robert L. 1954. *Project Mint Julep: Part III, snow studies.* SIPRE Technical Report No. 19. http://hdl.handle.net/11681/23588.

Shaw, Ian G.R. 2017. "The great war of enclosure: securing the skies." *Antipode* 49 (4): 883–906.

Turchetti, Simone and Peder Roberts (eds.). 2014. *The Surveillance Imperative: Geosciences during the Cold War and Beyond.* New York: Palgrave Macmillan.

Virilio, Paul. 1989. *War and Cinema: The logistics of perception.* New York: Verso.

Waterhouse, Robert W. 1955. *Structures for snow investigations on the Greenland Ice Cap* SIPRE Technical Report 27. http://hdl.handle.net/11681/5979.

Weiss, Eric D. 2011. "Cold War under the ice: The Army's bid for a long-range nuclear role, 1959–1963." *Journal of Cold War Studies* 3: 31–58.

Williams, Alison J. 2010. "A crisis in aerial sovereignty? Considering the implications of recent military violations of national airspace." *Area* 42 (1): 51–59.

5 Extractive spaces and the reproduction of remoteness

It is a characteristic strategy of mining companies that they will usually promote their projects by emphasising the uniqueness and scale of Greenland's geology and its mineral deposits. They are often described as being of the highest grade, and promises are made in the scoping documents and project strategies produced that mines will create economic opportunities and provide local employment. For example, Ironbark calls its Citronen Fjord zinc-lead project as located in an area that is "one of the world's largest undeveloped zinc-lead resources." Seeking to attract investors, the company released a Bankable Feasibility Study (BFS) in July 2021 that called it a "world class zinc project" in a "low-risk jurisdiction and emerging mining frontier." The BFS suggested there may be more mineral resources in the area than previously thought and it provided an assessment that the life of the mine could be extended to twenty years.[1] Bluejay Mining claims the area it is prospecting in for its Disko-Nuussuaq project has "potential to host mineralisation similar to the world's largest nickel/copper sulphide mine Norilsk-Talnakh ('Norilsk') in Siberia."[2] Bluejay also stresses the uniqueness of the ilmenite deposit at Moriusaq being developed by its Dundas Titanium subsidiary. Ilmenite is mined for its titanium – titanium dioxide is used as pigments in paint, plastics, enamels, paper, and in cosmetics and toothpaste, as well as in the making of a range of metal alloys. The Dundas Ilmenite Project has been classified as the world's highest grade mineral sand ilmenite deposit. And in much of its promotional material for developing the old mine workings at Maarmorilik, Black Angel Mining talks of there being a substantial exploration potential to locate new resources of a similar magnitude to what was extracted between 1973 and 1990.

All of the infrastructure being developed, as well as the technology and the transnational connections and labour markets the extractive sector relies on, the transnational practices and networks that are established, and the patterns of global consumption extraction depends upon have an environmental and social impact beyond that of the extraction of the resource. To say, as Greenland's political leaders do, that a turn to extracting critical minerals and a focus on sustainable mining is to ensure the development of a cleaner, green economy and help the world meet the challenge of climate change effectively is to ignore the reality that critical minerals needed for the production of electric vehicles or wind turbines, for instance, still need to be extracted and transported to manufacturing

DOI: 10.4324/9781003175421-6

facilities via the complex pattern of movement and flow that characterises the planetary mine.

Extractive industries present a significant challenge to how anthropologists approach and conceptualise the encounters between companies and the people who inhabit the places in which the hunt for oil or probing deep below the subsurface for minerals is taking hold (e.g. Larsen 2015). The formulations and configurations of frontiers and resource spaces, along with an array of abstractive, calculative, and speculative practices and the realignment of socio-economic relations, mean that extractive industries displace people and wildlife, disrupt and threaten biodiversity, pollute land, rivers, lakes, and seas, and destroy habitat and sacred sites (e.g. Fentiman 1996; Fentiman and Zabbey 2015; Karlsson 2011; Kirsch 2014; Li 2015). With reference to uranium mining and nuclear test sites in parts of Nevada, New Mexico, and California, environmental sociologist Valerie Kuletz (1998) shows how extractive spaces and toxic waste dumps that are situated in or close to Indigenous lands are viewed as being in empty wastelands. It is just one example of many that show how Indigenous perspectives on human-environment relations are too easily ignored, silenced, and subject to mechanisms of exclusion.

Anthropologists and other social scientists have catalogued cases from all over the world that detail how people and biodiversity are threatened by the environmental degradation caused by resource extraction projects (Jacka 2018). For example, Bax, Francesconi and Delgado (2019) show how the exceptional endemic species richness found in the Andes is being subjected to high rates of environmental destruction and natural resources exploitation. Similarly, Guzmán-Gallegos (2019: 53) remarks how in Peru's northern Amazonia region, oil development has "created landscapes of scattered debris." Elsewhere, Jacka (2015) explores how large-scale gold mining has polluted and degraded the environment of the Papua New Guinea highlands with consequences that are evident in the disruption of social relations, inequality, and violence; Li (ibid.) analyses the effects of mining on water, farmland, and sacred places, as well as what it means to live with toxic emissions in a smelter town in Peru; Fentiman (ibid.: 90) describes how oil has transformed the Niger Delta and has had significant social and economic effects on fishing communities since the first production activities and the construction of oil terminals in the late 1950s; and as Willow (2012, 2017) shows in her ethnography of the threat posed by industrial logging to the traditional territory of the Grassy Narrows First Nation in Ontario in Canada, extractive practices not only entail environmental degradation, they also threaten traditional land-based activities and affect social and cultural life.

Much of what anthropologists have written about in terms of these impacts is strikingly evident in the Arctic, where it has been well documented that mining and oil and gas activities have cumulative effects on traditional resource use practices, such as hunting, fishing, and reindeer herding, and on the economies and well-being of Indigenous and local communities (e.g. Chapin et al. 2005; Nuttall 2010). Tundra and boreal environments have been disturbed, often in violent ways, by extensive industrial development, such as oil extraction facilities, pipelines, and trails from seismic surveys, mining projects (abandoned mine sites have their own toxic legacies and controversies over the impacts on human health and the environment), or commercial forestry and clear-cut logging (e.g. Herrmann et al. 2014; Keeling and Sandlos 2015; Yakovleva 2011). Given this, there is a certain irony in how Dundas Titanium describes Greenland as "not only titanium rich, but also very beautiful." The company also states that the main conclusion drawn from the project's environmental impact assessment (EIA) report is that there are no material or unmanageable environmental impacts arising from the mine's development. The company's website points out how it is supported by the government and local community as highlighted on the EIA and social impact assessment (SIA), and reports how, in 2017, it was awarded "Prospector and Developer of the Year" by the Government of Greenland.[3] Local community support, as I discuss later, is not a view shared by all Inughuit hunters and fishers.

Rasmussen and Lund (2018) discuss how the making of frontiers and resource spaces and the territorialisation of resource control involves processes that, preceding any claim to legitimacy and assertion of authority, challenge, overturn, and replace existing patterns of spatial control, authority and social and institutional order. Frontier practices and extracting resources dissolve existing social orders, systems of property, and customary land rights, while territorialisation involves an erasure of place and a re-making and re-ordering of space. As the essays and case studies brought together in Cons and Eilenberg's edited volume *Frontier Assemblages* (2019) make clear, the processes of extraction and production redefine and incorporate margins and remote areas into new territorial formations and global networks, to which it may be added that these formations and networks are characteristic of the planetary in the sense that they are convoluted, menacing, contradictory, and fragmenting (Arboleda 2020). Resource frontiers are "sites of creative, if often ruinous, production," while "frontier assemblages" are "the intertwined materialities, actors, cultural logics, spatial dynamics, ecologies, and political economic processes that produce particular places as resource frontiers" (Cons and Eilenberg ibid.: 2). Cons and Eilenberg argue that:

> what matters in the incorporation (or re-incorporation) of margins are the various forces and processes that are assembled to reinvent these spaces as zones of opportunity. And second, we suggest that not only are these forces of spatial transformation resonant across sites, resources, and interventions, but that a broader view of territorial intervention gives us tools to understand a moment in which the relationship of millions of people to land and rule is being radically reconfigured.
>
> (Cons and Eilenberg ibid.)

Concerns are routinely expressed that the resource extraction projects being scoped out and developed in Greenland are in environmentally sensitive areas which are also (or are near to) cultural sites and vital places for hunting and fishing. Notwithstanding whatever local community support there may be for projects that companies outline in their EIA and SIA reports, there are public anxieties over social and economic impacts. Exploration activities inspire fraught political and public discussions concerning the future of Greenland, and doubts are expressed – Inuit Circumpolar Council-Greenland and local non-governmental organisations (NGOs) have been especially critical in this regard – whether a Greenlandic economy that includes an oil and mineral sector can be considered sustainable and proceed without major environmental disturbance and social disruption, no matter how much the consultants who produce assessments for companies like Dundas Titanium, for instance, insist that environmental and social impacts will be negligible and easily manageable. In places that would be close to or affected by extractive projects, many people express scepticism of such a claim and feel they are not sufficiently informed or consulted about the potential social and environmental impacts (e.g. Nuttall 2016), something to which I will return to discuss in more detail in Chapters 6 and 7. Before that, some discussion of the regulatory process and what is required in providing SIAs and EIAs for resource projects in Greenland is necessary. I then turn to discuss how ideas and representations of remoteness and distance are essential to some extractive projects, with specific reference to the Ironbark mine.

The politics of regulation

While mining companies may see climate change as opening up new Arctic maritime routes, making the production and shipping of what they unearth appear

Figure 5.1 Representations of remoteness matter for the construction of Greenland's resources, such as in Nuup Kangerlua (Photograph Mark Nuttall).

to be a less formidable logistical challenge, the ease of movement to and from spaces of extraction is not only about accessing icy waters to probe the seabed or crossing previously inaccessible terrain to dig deep into the subsurface as melting ice enables entry into remote places. Oil, gas, and mining companies wanting to operate in Greenland also encounter, negotiate, and work through the steps of a regulatory process that is often regarded by them – and, indeed, is often presented by the Greenlandic authorities – as being less bureaucratically compli-cated, restrictive, and lengthy than those found in other parts of the Arctic, such as northern Canada, which has public hearings and community consultation pro-cesses that companies often find to be frustrating in their logistical organisation and duration (e.g. Nuttall 2010). In Canada, the duty to consult with Indigenous groups has been affirmed by the Supreme Court in a number of decisions over re-source projects, including mining, hydropower, forestry, and road building. It is an important part of the federal government's activities, including a range of regula-tory project approvals at different stages, the licensing and authorisation of per-mits, operational decisions, policy development, and negotiations. Crucially, the duty to consult is about ensuring the recognition and protection of Indigenous treaty rights and land claims and where possible, acknowledging the importance to accommodate Indigenous communities whenever they contemplate decisions that may adversely impact asserted or established treaty rights. Consultation is also increasingly seen as a way through which relationships can be built between government, industry, and Indigenous people in Canada and a powerful process through which reconciliation can be advanced.[4]

Of course, the history of self-government in Greenland differs from Indigenous political movements in Canada in that there was no land claim settled in advance of the introduction of Home Rule in 1979, so consultation and negotiation do not proceed according to a set of procedures and standards that are guided by a recog-nition of Indigenous treaty rights or that members of Indigenous communities are beneficiaries of land claims. This downplaying of Indigenous identity and Indige-nous rights and even a diversity of Indigenous groups within Greenland with spe-cific identities was made apparent when, in 2003, the Inughuit who were moved to Qaanaaq in 1953 lost their case in the Danish High Court for compensation for loss of lands and the dispossession they suffered. The Kingdom of Denmark recognises that there is one Indigenous people in Greenland – the Greenland Inuit – or more accurately, it recognises that Greenlanders are a nation with the right to independence. This plays out in Nuuk in the way that Naalakkersuisut, the Government of Greenland, claims to represent the Greenlandic community as one that is homogenous and unified, with no divergent views or interests. To approach and consult with the government – or at least the Mineral Resources Authority – is to consult with Greenland as a single community. As Hannes Ger-hardt (2018) shows, the Government of Greenland and ICC-Greenland engage in different scalar constructions of identity, nationhood, and statehood. Each has different political agendas. ICC-Greenland emphasise Indigenous Inuit rights and argue for appropriate consultation with communities, while Naalakkersuisut is committed to a processes of nation-building and state formation, with the aim of eventual Greenlandic independence that does not privilege the idea of an Inuit homeland. The homepage of the Mineral Resources Authority makes an inviting statement that Greenland is "An underexplored, mineral-rich country with a competitive licensing framework, stable political environment, low investment risk and pro-mining population and government."[5] So, a company seeking information on mining prospects in Greenland will find everything they need from readily available geological databases, strategic impact assessments, oil spill sensitivity atlases, and surveys of biodiversity. The impression given is that there is no need to go to a community and have lengthy discussions to find out about the local knowledge of an area. The groundwork has already been laid out.

The origins of Greenland's current policy on extractive industries can be traced to the reform of mining legislation in the early 1990s. Extractive industries in Greenland are regulated by the Mineral Resources Act (MRA), which was officially adopted in 2009 and came into effect on 1 January 2010. The MRA provides a framework for the regulatory structure for mining and oil development and comprises a set of guidelines and international standards as reference points for potential investors and operating companies. It has been amended several times, most recently in November 2019.

A number of ministries and departments have various responsibilities for mineral resources. Greenland's government established the Bureau of Minerals and Petroleum (BMP) in 1998, which reported to the Minister for Industry and Mineral Resources. The MRA of 2009 established a one door policy for companies and granted the BMP exclusive power over the decision-making and control of extractive industry licensing and the management of activities. The idea was to allow for an expedited, streamlined approach, that would mean companies need only deal with one agency at all stages, from prospecting, to production, and to decommissioning at the end of a project's life (Hubbard 2013). Following an amendment to the MRA in 2013, the BMP was renamed the Mineral License and Safety Authority (MLSA). As was the case with the BMP, the MLSA is the administrative authority for licences, mineral resource activities, and licence-related safety matters, including the supervision and inspection of facilities and infrastructure. The MLSA sends out notifications, documents, and decisions to licensees and other parties covered by the MRA.

The other work carried out by the former BMP was distributed across several administrative units. The MLSA operates under the Ministry of Mineral Resources (MMR), which is responsible for general aspects of extractive industry regulation, but is concerned specifically with policymaking, legal issues, geographical considerations, and the marketing of mineral resources. It also exercises overall control over licensing for oil and gas exploration and production. The Environmental Agency for Mineral Resource Activities (EAMRA) has a mandate for overseeing all environmental aspects related to mineral resource activities. This includes environmental protection, issues relating to liability, and EIAs. A Department of Geology was also established within the MMR and carries out geological assessments, including potential economic evaluation of resources. Approval for exploration and production licences is the responsibility of Naalakkersuisut. The

Mineral Resources Authority – which operated under the Mineral Resources Act– is a collective term for all the government authorities, agencies, and departments that are responsible for mineral resources and mining. To make a point of clarification, and to avoid confusion, when I use the acronym MRA it refers to the Mineral Resources Act, not the Mineral Resources Authority.

Under the MRA, prospecting, exploration, and exploitation licences are only granted to public limited companies that are domiciled and registered in Greenland. In effect, this means that foreign companies establish Greenlandic subsidiaries, which they mostly entirely own, often locating their offices in Nuuk or other towns such as Ilulissat, Narsaq, and Qaqortoq – for example, Bluejay Mining's subsidiary for its project at Moriusaq is Dundas Titanium, which has its head office in Ilulissat. As I mentioned in Chapter 3, an increasing number of resource projects are currently going through prospecting and exploratory phases as well as the approvals process for exploitation and operational licences.

Public participation and public consultation processes

According to the MRA, companies must provide an EIA for public consultation and government approval. The EIA process has to be integrated into the exploration phase of a project, but its general purpose is to identify and consider the environmental issues and the potential environmental impacts that will need to be addressed and resolved throughout all project stages. As well as outlining the steps for preparing an EIA, project proponents have to fulfil all its basic requirements, including environmental baseline studies, disturbance aspects, local use of the area, and local knowledge studies, noise, dust, and other environmental considerations. The reports they produce have to be available in Greenlandic, Danish, and English.

Public participation is required within the EIA and SIA processes in Greenland. Following the completion and submission of the appropriate documents, companies are expected to hold public meetings and consultations in all communities that would be affected by a project. A pre-consultation phase is now key to this process and a project description must be submitted for public pre-hearing for 35 days before the consultations are held for the EIA and SIA. Partly a response to critiques of the absence of strong public involvement (e.g. see Ackrén 2016; Nuttall 2013), the purpose of the pre-consultation hearing phase is to make things more transparent by allowing the Greenlandic public an opportunity to participate in the discussions surrounding a project. In theory, it allows people an involvement in the process by giving them time to be consulted before major decisions over a project are made. This is usually the first opportunity for the public to receive information about an extractive industry project, to provide their comments and come forward with concerns and suggestions. The comments received must be included in the company's overall planning process for the project.

The duration, content, or procedures of these meetings and consultations are not specifically defined. Yet the outcome of the meetings and consultations should lead to an appropriate amendment of an application, especially if public concerns are raised and taken into account. However, there is currently no requirement that the consultation process itself needs to be considered during the application or approval processes. Public response is used by companies to refine and amend their documents, but there is no requirement that the proponent should state why it did not make any changes that may have been suggested during the consultation process. So, the situation is that the production of both an EIA and SIA in Greenland is a legal requirement for companies in their planning process and are the only documents needed by decision-makers when considering the potential environmental and social impacts a project may have. There is no consideration of any transcripts of public meetings, and the concerns and questions that may have been raised, that may have been produced, and which are a matter of public record. If an EIA or SIA that a company produces says there is community support for a project and if there one or two quotes from local residents to back this up, then usually this is all that is needed on the part of the agencies that comprise the Mineral Resources Authority.

The government's guidelines on preparing an SIA were first published in 2009, but were revised in 2016. They now begin with this statement:

> Greenland wants to develop the mineral resources area into one of the country's primary and principal business sectors. This is to be done in close collaboration and dialogue with the Greenlandic population. Understanding the interaction between mineral projects and Greenlandic society is therefore essential in order to create sustainable relations between mineral resources companies, municipalities, affected individuals, other stakeholders as well as Greenlandic society in general.
>
> (Government of Greenland 2016: 4)

Despite this declaration of the importance of companies and their consultants working in close collaboration with local communities, exploration activities continue to provoke public concern and inspire political debate. ICC-Greenland and local NGOs have been especially critical of the regulatory process and argue that the consideration and assessment of social, cultural, and environmental impacts are not necessarily robust enough. It is important to note in this regard that the SIA guidelines give no definition or explanation of what 'consultation' is and how it should proceed. The guidelines do, however, make this reference to traditional and local knowledge:

> When preparing the SIA report, the company must use traditional and local knowledge as far as possible by collecting information through qualitative interviews. It is important to incorporate local knowledge from individuals, commercial hunters and fishermen etc. This knowledge may have been passed on from generation to generation and has not necessarily been described and analysed in publications and public literature. It is recommended that the licensee describes, analyses and uses the traditional knowledge existing in the area in the SIA report. This knowledge may also include municipal planning documents and similar descriptions.
>
> (Government of Greenland ibid.: 24)

Yet, again the guidelines offer no suggestions nor do they set out ways of think-ing for how local or traditional knowledge is to be analysed and incorporated, and there is no reference to recognising free, prior, and informed consent, which is, according to the UN Declaration on the Rights of Indigenous Peoples (UNDRIP), a prerequisite for engagement with Indigenous communities. In theory, at least, consultation is supposed to offer an inclusive, equitable process of discussion and dialogue, one that enables Indigenous people to participate in an assessment pro-cess when extraction projects with significant socio-environmental impacts are planned and for each aspect of the project, including the technical material and reports, to be evaluated critically as well as for public concern and suggestions to be considered by the proponents. The SIA guidelines merely recommend, rather than require, that licensees describe, analyse, and use traditional knowledge.

I would argue that part of the problem is the conflation of 'public meetings' and 'consultation.' Public meetings are advertised by the Mineral Resources Au-thority as opportunities for the public to receive information about a project from the company as proponent as well as to ask relevant questions; meaningful con-sultation, however, implies a deeper engagement with communities concerning the issues surrounding a project, its technical components, its potential social and environmental consequences, and any concerns people may have about the potential impacts on hunting and fishing activities. This appears to be lacking, however. In my experience of attending public meetings and information sessions about projects – meetings that are held to do just that, to provide information – companies often refer to them as public consultations without recognising that there is a significant difference between holding a 'town hall' style information meeting and engaging in consultation.

As Flemmer and Schilling-Vacaflor (2016) point out, the right to prior con-sultation and to informed consent represents the basis of a new global model that is shaping state-Indigenous relations; but drawing on research on the hy-drocarbon sectors in Bolivia and Peru, they describe the difficulties Indigenous groups have in defending or articulating their concerns, demands, and even their visions. Some of these difficulties arise from asymmetries in power structures, problems with effective communication at meetings with resource companies and government officials (including translation of scientific concepts and Indigenous knowledge), timing, and simplistic assumptions that underlie the consultation ap-proach. This latter point reinforces my argument about the way 'public meeting' and 'consultation' are interchangeable in the Greenlandic context. For Flemmer and Schilling-Vacaflor, this leads to a situation where the new model leaves prom-ises that are unfulfilled.

In Greenland, Dahl and Hansen (2019: 184) argue that while there seems to be a general expectation in the impact assessment guidelines that Indigenous knowl-edge will be included, there is nonetheless a lack of experience – both in the com-panies working in Greenland and with the consultants they employ – in how to reflect on it, incorporate, and utilise it in an appropriate manner once it has been brought into the assessment process. The value of Indigenous and local knowl-edge is also often downplayed. I recall a conversation I had with a Nuuk-based consultant (who was originally from Denmark) in late 2019 who expressed scepticism about the need to consider and use Indigenous and local knowledge in project assessments: "People just talk about seals and fish," he said, "but this isn't the scientific data you need. You have to have facts and evidence and things you can test with proper methods." He went on to explain that reports have to be robust enough to stand up to scrutiny and that local knowledge amounted only to comments, statements, and opinions that could not be reliable. And on another occasion, in 2011, I met with a consultant from the UK who was tasked with carrying out an SIA for an oil exploration project in Southwest Greenland's waters. He had a background and career history in engineering and was visiting Nuuk for five days – it was his first time in Greenland, and he had no linguistic skills in either Greenlandic or Danish – and, explaining to me that he was under some pressure to learn about Greenlandic society and culture, asked if I could, in his words, "recommend some books" that would be useful for his assessment. I suggested to him that it was probably a good approach if he was willing to take the time and care to get to know something about Greenland's history, society, and politics through extensive reading and research, but that it would be a worthwhile investment if he spent a long period talking with people about their culture, listening to them express their feelings about community, their relations with animals and their surroundings, and about their views on oil exploration.

While companies – or usually the consultants they hire – are expected to carry out baseline studies and environmental and SIAs in accordance with Greenland government regulations, they usually proceed to do so without any particular definition of the environment or the social world that is appropriate for local contexts. Indeed, in my experience with talking to company representatives and consultants, there seems to be an implicit assumption that those who carry out such work know what 'environment' and 'social' mean, even if they do not take too much time to learn about Greenlandic history and society. The things making up, filling, and constituting an environment or society seem apparent and observable to them and the potential impacts and benefits measurable and quantifiable. It seems to be a matter of describing how environments and societies are sensitive, vulnerable, and likely to be impacted by a mine or oil extraction. All consultants have to do is assess the impacts and propose mitigation and monitoring initiatives. In short, things appear self-evident to consultants who fall back on technocentric language that describes environment and society in terms of 'systems,' while measures to avoid or mitigate risks and impacts are set out in reports that are submitted to companies and those government departments concerned with approving and regulating extractive industry.

One overall aim of an EIA, for example – and this seems obvious enough, of course – is to identify, evaluate, and minimise the environmental effects of a project. However, the purpose of an EIA carried out in Greenland's resource spaces is also to provide Greenlandic authorities and other stakeholders with information on the environmental impacts of the project, including how the mining company intends to promote environmentally sound and sustainable solutions through the identification and implementation of mitigating measures. Greenland's guidelines for EIAs say that they must contain "a thorough description of the state of the environment before the start-up of mining activities." But there is no fieldwork carried out as part of an EIA. Essentially, EIAs are baseline, desk studies. They allow certain people – consultants with expertise, an array of scientists – to speak for and describe how the world is and how it is likely or not to be affected by a project.

An EIA may describe the ecologies, biota, and physical characteristics of potential resource spaces – drawing from published and available scientific literature and reports – including some detail of the socio-economics and demographics of the area under assessment. Yet, water and ice, land, animals, and fish resist easy definition, description, and classification. For example, as I discussed in Chapter 4 and will return to in Chapters 6 and 7, understanding sea ice, glaciers, icebergs, and oceanic troughs and ridges in northern Greenland has involved a considerable investigative labour involving military and scientific personnel from the Cold War to the present, with the aim of sizing up, measuring, and making the region and its non-human entities known and legible. This is often at odds with Indigenous understandings and accounts of people's surroundings and of human-environment relations. The environmental and the social co-constitute one another in profound ways, but the slippery and somewhat stretchable nature of northern Greenland – the elusive manner of animals and fish and how Inuit living there relate to their surroundings and engage with the non-human – has often made any description and accounting of how the shifting, dynamic, and unpredictable elements of the region intermingle and interact an extremely difficult task (Dodds and Nuttall 2019). Given this, the intricacies of the environment and the relations between the human and non-human are not captured in an EIA or SIA.

Like EIAs, many of the SIAs carried out in Greenland are based on secondary sources and, as I have already mentioned, are often primarily desk studies. For example, the methodology section of the draft SIA for the Dundas Ilmenite project at Moriusaq states that:

> Most of the baseline information presented in this SIA is based on information available from secondary sources. The sources include research reports, relevant studies, official strategies and statistical data from Statistics Greenland. Efforts have been put into presenting the most updated information at the point of writing. At local level, some primary sources have been used to describe the baseline situation. In the baseline it is indicated when information is received from primary sources.

The SIA goes on to explain that data collected from primary sources was done so for the purpose of qualifying the data collected through the secondary sources. In-person interviews and meetings were held during trips to Qaanaaq in February 2017 and Nuuk and Ilulissat in October 2017. Savissivik, perhaps the community to be affected the most because of the potential impacts of traditional hunting areas, was not visited. On visits I made to Savissivik between 2014 and 2017, one of the major concerns people there repeatedly told me they had was the increased presence of mining companies in the area, especially in and around their customary hunting places and travel routes, but that they felt they were not given information about potential projects and the prospecting that was going on.

In March 2015, for example, Lene Kielsen Holm and I spent time in Savissivik to carry out interviews and have discussions with people there about their experiences of, and perceptions about, changing sea ice, as well as some of the barriers they encountered when seeking ways of anticipating and responding to the transformations in their surroundings. One evening, we held an open discussion session in the community hall and talk soon turned to concerns that the impacts of climate change on their livelihoods would be exacerbated by seismic activities as part of oil exploration campaigns, especially the possible impacts on narwhals (see Chapter 6, for more on these seismic surveys). But people also wanted to talk about the fact that there was a paucity of information about other activities and the conversation moved to mineral prospecting. Often, they did not even know what work was going on – they would see helicopters, boats, and the camps of geologists, but had not received notification of these activities.

When a draft EIA/SIA report has been approved as fulfilling the minimum requirements as far the regulatory process is concerned, it is published on the Government of Greenland website for an eight-week period of public consultation. During this time, the project proponent is required to hold meetings in the towns and settlements that would be affected by the project. The public consultation period allows people to submit comments in oral form at the meetings or as written submissions, offering another opportunity for traditional and local knowledge to be incorporated (Dahl and Hansen ibid.). Following the meetings and the submission of comments, the project proponent revises the draft EIA/SIA and resubmits it to the Government of Greenland along with a white paper for review and, if acceptable, approval, which is necessary for submission of an exploitation application licence. In theory, the public consultation period allows an opportunity for people to express their opinions, views, and concerns, but they often encounter difficulties and obstacles in doing so. The EIA and SIA reports for the ilmenite project were published on the government's website, along with links to supporting documents on the Dundas Titanium website. However, people in Savissivik and Qaanaaq often encounter difficulty with a reliable Internet connection; so accessing, downloading, and responding to the assessments, various reports, and supporting documents is not always an easy task. Furthermore, the Covid-19 pandemic made organising the consultation meetings difficult. Travel restrictions to and within Greenland meant that representatives from Dundas Titanium and consultants who were not resident in Greenland could only take part in the consultation meetings online.

The SIA methodology for the ilmenite project also included telephone and in-person interviews with researchers and key stakeholders between December 2018 and February 2019. Overall, the draft SIA is favourable about the benefits of the project, specifically citing 270 jobs for Greenlandic employees. However, it also highlights that the project will lead to restricted public access to the licence area, which is used by local hunters, but it concludes that the negative social impacts will be relatively small and easily mitigated. This conclusion is made despite there being no opportunity for hunters to talk about their travel routes, the places of importance for hunting and fishing, or any other observations they could have made about animals, breeding sites, ice conditions, and so on. The SIA contains summaries of responses and comments from people in Qaanaaq who attended the open information meeting in February 2017 as well as from semi-structured interviews with eight people in the community (four of the interviewees were hunters; Qaanaaq has a population of around 650 – about 50 people are full-time or occupational hunters, so four are not necessarily representative of the wider community). The report states that there are many in the community who thought the project would bring some economic benefit and employment, but there were many who also expressed concern over its impacts on the environment and on hunting and fishing activities. Yet traditional knowledge or local knowledge is not incorporated in either the draft SIA or the draft EIA for the project – rather, it appears that, for consultants, local knowledge equates a response, concern or opinion.

Abstractions and frontiers

As I have shown so far, growth-oriented imaginaries are nothing new for thinking about Greenland's future, as fish and minerals (and even sheep) became the basis for plans for modernising and industrialising the country at various times in the twentieth century (e.g. see also Priebe 2017). However, one of my main arguments in this book is that in the last decade or so, a representational shift has become far more apparent in development discourse concerning Greenland's resource potential. In this discourse, Greenland is no longer viewed as a far-flung Arctic territory with a harsh, unforgiving environment that hinders resource exploration and extraction. It is framed as a dynamic, globalising country that is open for business and made increasingly accessible, not just by melting ice but by a governing élite that welcomes extractive industries, even if oil exploration has been suspended for the time being. Yet, the remote and distant are entangled with the ideas and ambitions for the building of a prosperous, globally connected nation.

To be a major hub in a Global Arctic, Greenland needs to create its own peripheries and mark out its own sacrifice zones to enable extraction. Accompanying this representational shift, as I have observed it, is a focus on investment in the technological and engineering expertise that allows (and celebrates) extraction and exploitation in wild regions, with technoscientific inscriptions marking out lands and waters as resource zones that can be mapped, surveyed, tamed, and enclosed, but which still remain at enough of a distance from population centres and vital ecosystems, companies argue, so that social and environmental harm will be minimised.

Gavin Bridge (2009a) suggests that we should think of the world's geographical spaces of extraction as akin to portals in which time and space work differently. There is a world of 'natural production' underground, he argues, a place within the Earth where the processes of deep time lead to the creation of the concentrations of hydrocarbons that we call fossil fuels or the chemical compounds we know as minerals. Above ground there is a world of 'social production,' of mobility,

Figure 5.2 Sikuki is Nuuk's growing sea port and is essential for Greenland's greater global connectivity (Photograph Mark Nuttall).

transformation, and distribution, where what are defined as raw materials are converted into things that have value and enter global supply chains – the human appraisal of the things of the underground, as Erich Zimmermann articulated it. For Bridge, a mine shaft or an oil well is a hole in the ground, a point of access to subsurface resources. First of all, this point of access needs to be located and marked. Following this, exclusive control over it needs to be secured. All this involves political processes that raise questions over land use and power, social dispossession and deterritorialisation, but the point at which hydrocarbons or minerals exit the ground and reach the surface is the beginning of an energy commodity chain. Extractive industry, then, also involves technologies and infrastructures that are characterised by transportation, flows, and mobilities over great distances of land and sea, the production of energy, and the manufacture of goods from oil and minerals, as well as economic and financial networks, markets, and other patterns of work and employment, that stretch around the globe (e.g. Arboleda ibid.; Hein 2018).

Espig and de Rijke (2018) argue that global energy networks – and to this can be added energy commodity chains – prompt anthropologists to rethink the nature of their 'field sites' and how they can pay attention to and understand extractive industries and their wider reach and social, economic, and environmental impacts in relation to global interconnectedness. As Hein (ibid.) puts it, constellations of oil actors, which include corporations and nations, shape the physical spaces that appear to be disconnected and geographically distant, but which emerge as a global palimpsestic petroleumscape (see also Ferguson 2005). In a similar vein, and drawing on research in West Africa, Schritt and Behrends (2018: 212) suggest that oil extraction is located in 'oil zones' that are trans-territorial spaces. They point to the:

> (dis)connections between (distant) places through infrastructures, technologies, borders, fences and security practices, as well as discursive movements of ideas, theories, models and narratives. The concept of "oil zones" thus enables us to look at specific places in Niger and Chad, like the oil extraction sites that are disconnected from their immediate surroundings by fences and military guards, but at the same time connected with distant places in Africa and beyond (United States, Europe, China) through pipelines, specific means of transport and the conversion of crude oil into new products. It also helps us to analyse the particular and quite different forms of (dis)connection between places on a global scale in relation to the ideas, scientific theories and economic models that bring forth these very (dis)connections.

This planetary scale of extractive industry means that territories of extraction, the places that are dug into to get the 'neutral stuff' unearthed so it can be turned into resources (cf. Zimmermann 1951), become intertwined with other elements and fragments of extractive industry (Arboleda ibid.). Mazen Labban's conceptualisation of the planetary mine draws attention to understanding extractive industry as something that transcends a bounded geographical location and the productive activity of getting ores, metals, and hydrocarbons out of the ground. Once they emerge from the hole in the ground (cf. Bridge ibid.), they have to move in global and planetary space and so "the planetary mine comes into being in the transformative circulation of materials between places" (Labban 2014: 564). Martín Arboleda builds on Labban's notion in his discussion of mining in Chile's Atacama Desert and its global reach, to show how geographies of extraction are "entangled in a global apparatus of production and exchange that supersedes the premises and internal dynamics of a proverbial world system of cores and peripheries defined exclusively by national borders" (Arboleda ibid.: 5). The point here is that a mine is not a "discrete sociotechnical object but a dense network of territorial infrastructures and spatial technologies widely distributed across space" (ibid.). The same can be said of an oil or gas field.

Extractive industry – even if it is focused on rare earth elements and critical minerals used in cleaner energy – requires a vast infrastructure and associated technologies as well as the bureaucracies of finance, sales, and export beyond a mine or oilfield and extending beyond encampments, boomtowns, and the communities that grow up around extraction. Once coal or iron ore, for instance, have been extracted from the ground, these materials have to be transported via railways and roads or by way of harbours and ships to terminals, production, and processing facilities and on to storage facilities and markets. Oil and gas production includes wells, rigs, gathering facilities, pipelines, pump stations, processing plants, and refineries.

Pipelines are good examples of this infrastructural apparatus and the "complex infrastructures of connectivity" (Arboleda ibid.: 16). They are not to be understood merely as single lines of steel pipe that are buried underground, and so out of view, running from a place of extractive origin delivering oil or gas to a terminus. They are components of these vast, complex, global technological systems that make up the planetary nature of extraction, production, and consumption. Writing about the US-Canada natural gas industry, Arthur Mason (2012: 80) describes it as a "large technical system consisting of a vast continental-sized machine made up of 2 million miles of steel pipe." Because this complex, interconnected system of pipelines is buried and unseen, Mason says that only a few people are aware of, experience or witness the scale of the convoluted techno-ontological dimension of this network first hand; but he argues that "the steady stream of fuel to consumers and occasional explosion that destroys lives and entire neighbourhoods" is testimony to the very real material and political presence and the impacts pipelines have (ibid.). Hein (ibid.) explores one dimension of this with reference to the relationship between oil refineries and port cities, their entanglement with global events and national strategies, and their spatial and environmental impacts. Yet, as Stuart Kirsch argues, given how large-scale resource extraction projects are more often than not dominated by international corporations, influenced by distant capital, and are primarily responsive to global markets, mining companies are discouraged from "making sufficient investments in environmental controls" (Kirsch ibid.: 15). If the sheer scale and nature of the infrastructures of connectivity are difficult to experience and measure and if public awareness of them is dimmed once pipelines, for instance, are buried and out of sight, then the need for accountability is perhaps not necessarily always a priority.

While Bridge talks of two worlds in which time and space work differently – the underground world of natural production and the surface world of social production – the concentrations of hydrocarbons and chemical compounds that have been formed, fashioned, and shaped in deep time must first become resources before holes in the ground are dug to extract them. The environmental footprint and the social and economic impacts of mining and oil and gas production are not just confined to geographically bounded locations that have been marked out as resource zones within countries. But neither can resource spaces be seen just as territories in which extraction from the earth occurs. There is a middle world between the underground and the surface, a world of abstraction that gives form and value to organic and chemical compounds that allow them to become fossil fuels and minerals (see Mason 2022). As I reflected on in Chapter 3, within as well as beyond the productive activities of oil, gas, or mining, extractive industry involves abstraction and calculation and the formation of frontiers and resource zones through the investment, work, and labour of companies and their retinue of consultants, geologists, geospatial analysists, engineers, and economists. They produce data, maps, assessments, and a wide range of other documents that give form and meaning to spaces of extraction.

Michael Watts (2015: 215) points out that frontiers are "time-spaces endowed with quite specific sorts of properties and qualities" by which the conditions for extractive accumulation are put in place. However, to extend this by drawing on Alfred Whitehead's (2007: 85) phrasing, impact assessments, regulatory frameworks, and governance structures for extractive industry seldom, if ever, concern themselves with how "abstractive elements form the fundamental element of space and time." In Greenland, as in any resource territory, abstraction includes the assembling of the scientific, technical, and economic material as well as speculation on temporal and spatial scales that are necessary for resources to become. Abstraction involves geographical imaginaries that are evoked and sustained by geophysical probing and the cartographic production of both the surface and subsurface (as well as oceanic environments, including the seabed), all of which emphasise an aesthetics of geological strata, minerals, and hydrocarbons that are suggestive of economic viability and profit. Abstraction allows for the exposure of geological formations and mineral veins through various processes of visualisation and through a politics concerned with the promise and potentialities of the subterranean.

In writing about how frontiers are formed, re-created, pushed back, and exploited, anthropologists have been concerned with understanding how places in which resources are defined, located, made, and extracted are imagined and classified as remote and pristine, empty and wild. In this way, they become subject to techniques and practices of conquest, abstraction, commodification, and export-oriented resource extraction. The visualisation and reproduction of remoteness is essential to this abstractive process. Anna Tsing argues that:

> …frontiers create wildness so that some—and not others—may reap its rewards. Frontiers are unregulated because they arise in the interstitial places made by collaborations among legitimate and illegitimate partners: armies and bandits; gangsters and corporations; builders and despoilers. They confuse the boundaries of law and theft, governance and violence, use and destruction.
>
> (Tsing 2005: 27)

In her account of logging and mining in the rainforests of Kalimantan, the frontier is what Tsing considers a "zone of awkward engagement." The extractive frontier, she argues, is an imaginative project in a site of transformations that are "made in the shifting terrain between legality and illegality, public and private ownership, brutal rape and passionate charisma, ethnic collaboration and hostility, violence and law, restoration and extermination" (Tsing ibid.: 33). Similarly, Mary Louise Pratt used the term "contact zones" to describe those spaces where "cultures, meet, clash and grapple with each other, often in contexts of highly asymmetrical relations of power such as colonialism, slavery, or their aftermaths as they are lived out in many parts of the world today" (Pratt 1992: 4).

James Ferguson's (ibid.) discussion on extractive enclaving, a process that he argues entails little or no economic, social, or cultural benefit for the places in which extraction occurs or for the wider society, provides anthropologists with a conceptual framework for understanding how extractive practices impact the daily lives and surroundings of people who live not so much on extractive frontiers or zones, but in extractive enclaves. The extractive enclave arises from extractive practices that aim to maximise profit, whatever the social and environmental costs may be. Discussing various contestations over the ownership of oil in the Niger Delta and how local communities do not benefit, Omolade Adunbi (2020) points out that the extractive enclave is shaped by a global order that connects local and national actors with transnational practices and networks that enable the emergence of alliances and arrangements that privilege the state, oil corporations, and business initiatives over local communities and environments.

Resource spaces are such contact zones of encounters, antagonisms, contestations, struggles, and entanglement (Bainton and Owen 2019; Bridge 2004; Pijpers 2019). As Gómez-Barris (2017) remarks, the extractive zone is where coordinated forms of capitalist power, technologies, and worldview mark out and reduce places to spaces of digging out, production, processing, and transformation in which natural resources are made and converted into global commodities. In the rest of this chapter, I show how this is so with reference to Greenland's far northern mine.

The far northern mine

In Greenland's remotest regions, international mining companies are engaged in the prospecting of extractive terrain and are assembling resource spaces as they develop plans for a number of extractive ventures. I have already mentioned that Ironbark Zinc Limited, an Australian company, is developing a large zinc-lead mine project at Citronen Fjord. The fjord is actually a small bay of around 4 kilometres long on the southern side of Frederick E. Hyde Fjord in Peary Land on Greenland's far northern, uninhabited edge. The zinc-lead mine is also the world's northernmost commercial project and is in an area that has a large sulphide deposit (Andersen 2001; Jensen and Thorning 2001). My use of 'remotest,' 'far northern,' 'uninhabited edge,' and 'world's northernmost' is purposeful here, for this is consistent with how North and Northeast Greenland have been described in the annals of exploration, adventure, and geological research as well as by mining companies operating in the country today. I discussed aspects of this in Chapter 1. It is also consistent with how the making of resource sites and mineral and hydrocarbon reserves involves cultural, economic, and political dynamics of frontier-making, territorialisation, and deterritorialisation, all of which are processes that threaten, subvert, and dissolve existing social worlds and re-order surface and subterranean spaces (Günel 2019; Jacka 2015; Rasmussen and Lund 2018).[6]

I began Chapter 1 with a consideration of how Greenland has been imagined, perceived, and written about as a remote Arctic space. Gazing upon remoteness – and entering into remote places – is often associated with personal transformation. Approaching the west coast of Greenland by ship in summer 1932, for example, glaciologist Ernst Sorge and his intrepid colleagues felt they had reached the Arctic once they had sighted their first iceberg. Sorge had a previous experience of Greenland, notably as a member of the German Greenland Expedition of 1930–1931 which was led by Alfred Wegener, and measured climate conditions and the dynamics of the inland ice. His journey in 1932 was as a member of an expedition led by Arnold Fanck, a German producer of films in mountain regions who wanted to portray the life of icebergs, but also produce a drama based on Wegener's expedition. The result, *S.O.S. Eisberg*, was filmed in the Uummannaq district and was released in 1933. Sorge's narrative about the journey, the filmmaking process, and the ensemble of personnel and characters involved, including the actors, is called *With 'Plane, Boat and Camera in Greenland*, and it was published in 1935. The encounter with that first iceberg marked "the real beginning of our expedition," he wrote in the book. "From now on we got the notion we were really off Greenland, and had become polar adventurers. We allowed our beards to grow so as to penetrate these regions in appropriate and professional sort of seeming" (Sorge 1935: 31).

Sorge describes an acting out and display of the appropriate postures and appearances that seem necessary when 'entering' and 'penetrating' the Arctic, and this is consistent with an enduring trope. Exploration and mapping the unknown were often regarded and written about "as the *conquest* of virgin territory" with the Arctic perceived "as a particularly masculine arena" (Bergmann 1993: 53), an empty space where masculine identities could be expressed and tested (Hill 2008). I discussed how this was so with reference to the Danish exploration of Northeast Greenland in Chapter 1, but it had a pervasive influence on Arctic journeying and discovery through the circumpolar north. The nineteenth-century Arctic explorer, in particular, was portrayed in exploration narratives, literature, and poetry as someone who displayed "a dogged perseverance through enormous difficulties" in a vast empty space where they would be faced with horror, disaster, or triumph (Behrisch 2003: 83). If gazing upon, approaching, entering, and setting about the exploration of a remote place are associated with personal transformation, then once those intrepid adventurers get there, the place itself is unavoidably transformed by human presence through various acts of claiming, colonisation, and narrative depiction that influences the way others see it, regard it, and treat it.

One of Edwin Ardener's (1987) insights is that encounters with remoteness are often accompanied by a sense of loss as imaginary places are made real. The paradox, he suggests, is that travelling to a remote place is to be aware of its accessibility, and hence its vulnerability and exposure to forces of disruption and change. Resource exploration today has more of a sense of directed searching in places that promise to hold 'undiscovered' but 'recoverable resources.' In this way, exploration does not necessarily entail a process of personal transformation of the explorer (although this is not to deny that geologists must find working in Greenland quite exhilarating), but a more targeted probing that will transform remote places into extractive resource spaces. In Greenland's northernmost reaches, remoteness, geology, and geomorphology along with logistical challenges (such as a lack of airfields and harbours) and the presence of sea ice have been cited as impediments to exploration in recent decades (e.g. Jensen and Thorning ibid.). The early geologist was also seen as a pioneer, an explorer – just as Lauge Koch was in northern Greenland – and was characterised as someone who demonstrated perseverance in difficult conditions and remote terrain. Today, however, companies such as Ironbark argue that advances in mapping and surveying techniques, technology, and shipping as well as the effects of climate change on sea ice mean there are fewer obstacles to exploration, extraction, and transport. The area is no longer difficult to penetrate, but the construction of the Arctic as remote space persists in the texts and documents produced for and by resource exploration companies, just as it characterised the narratives of early explorers in the nineteenth and twentieth centuries (Hanrahan 2017). Exploration is now focused on discovering and locating natural resources, but the view that there is greater accessibility because of a combination of advances in technology, new mapping techniques, and melting ice suggests that speculation is more a matter of certainty in that resources will be found, can be extracted from remote places, and can be transported from them.

There is a long history of geological research and resource speculation that has focused on northern Greenland and Peary Land, and this has aimed to map and fill in the blank spaces. Lauge Koch first reported on preliminary geological work carried out in northern Greenland, including north of the Peary Channel, during the Second Thule Expedition led by Knud Rasmussen in 1916–1918, and he returned in the 1920s and 1930s to do more exploration of the region. In 1922, Lauge Koch also published an article in *American Anthropologist* on ethnographic observations on the lifeways and culture of Inughuit he encountered and spent time with on the southern coast of Washington Land. When Koch began his topographical and geological mapping, knowledge of the geography of northern Greenland was imprecise and the coastal areas were sketched out only in broad outline. Parts of the northwest had been described and mapped mainly during nineteenth-century British and American expeditions – which were usually on their way to somewhere else in the Arctic, such as to find the Northwest Passage or a route to the North Pole – and later by Robert Peary in the 1890s and first decade of the twentieth century. Peary also travelled around the northernmost stretches of Greenland at Kap Morris Jesup, while parts of Northeast Greenland were explored during the Danmark Expedition of 1906–1908, the Alabama Expedition of 1909–1912, and the First Thule Expedition of 1912 (Dawes 1991). The inland areas north of 80° and the nature and extent of the region's fjords were not known. On his early expeditions, Koch travelled extensively by dog sledge, but he also paved the way for modern forms of exploration and mapping (Ries 2002).

Other geological and topographic mapping activities took place during the Danish Peary Land expedition in the 1940s, while William E. Davies carried out extensive mapping during expeditions sponsored by the US military in the late 1950s and early 1960s to determine the nature of Greenlandic terrain. This is an area along with the northern part of Canada's Arctic islands that is underlain by the Paleozoic Franklinian Basin. It is a significant geological formation that has long been the target of exploration for evidence of lead-zinc mineralisation and the presence of oil. The large-scale surveying of the region and its resource potential began in the 1970s (Andersen ibid.). Platinova-Nanisivik Mines undertook a reconnaissance of Northwest Greenland in 1992, while exploration and assessment of the Citronen Fjord area's lead and zinc deposits has been carried out since the early 1990s. This has involved a penetration of the area's depths rather than just the mapping of its surface layers, and activities have been carried out by several companies with interests in the region's potential. Platinova, for example, discovered a sulphide deposit with lead and zinc in 1993 and went on to map, survey, and drill there over summer seasons until 1997. The mineralisation is in permafrost. The data from Platinova's detailed geological mapping, together with that gathered from research funded by the Danish Research Councils on resources in the sedimentary basins of North and East Greenland (Kragh, Jensen and Fougt 1997; Langdahl and Elberling 1997), informed an economic assessment of the exploitation aspects of the sulphide deposit (van der Stijl and Mosher 1998).

Scientific baseline studies were carried out for the EIA, and in anticipation of development, the Danish Ministry of Environment and Energy's National Environment Research Institute produced a report assessing the possible environmental impacts of shipping to and from Citronen Fjord (Boertmann 1996). This assessment was a cautious one. It noted that there was little information available at the time on the biological and oceanographic characteristics of the region's marine environment. As no resource exploitation was then occurring and the size and type of vessels to be used in the eventual transit of mined ore was unknown, the assessment concluded that it was not possible to determine the likely effects. The ministry's report did though point out the sensitivities and vulnerabilities of the Northeast Greenland marine environment – given that this is also a national park – and highlighted key areas of wildlife habitat. It noted that incidental oil spills caused by ruptured tanks probably posed the most serious threat to the environment along the sailing routes to Citronen Fjord, and it recommended that the Northeast Water Polynya (and icebreaking through the ice barriers to the north and south of it) should be avoided.

At the beginning of July 1994, a team from the Greenland National Museum in Nuuk visited Citronen Fjord for four days to conduct an archaeological assessment of the area where Platinova was carrying out its exploration and drilling. The survey report described Peary Land as a marginal area and concluded that despite evidence of past human occupation, there were no major Paleo-Inuit sites in the area that would be affected by Platinova's activities. Indeed, as the report's author concluded, the 'character' of the finds was called into question as informing an assessment of them as 'archaeological sites' that would have been protected by Greenland's conservation act (Kapel 1994). Many archaeologists who specialise in northern Greenland would likely disagree about the paucity of evidence of Inuit occupancy there and how sites are classified and categorised. For one thing, Danish Arctic explorer and archaeologist Eigil Knuth, who lived from 1903 to 1996, carried out extensive surveys and excavations for over six decades in Peary Land and other parts of the north and northeast of Greenland's coastal and interior ice-free regions which contributed enormously to the understanding of 'the northernmost ruins of the globe' (Grønnow and Jensen 2003). Yet, by describing the traces of past human settlement in such a remote, marginal area in the way the survey report did, the history of human dwelling was itself marginalised and considered as not so significant. Platinova did not move forward with plans for developing the area, but other companies continued to express interest in the resource potential of Citronen Fjord.

In December 2016, Greenland's government granted Ironbark a production licence to mine three deposits in the area. This licence allows both open pit and underground mining operations, with an on-site processing facility to produce zinc concentrate for shipment to third party smelters for refining. Impact assessments, including archaeological surveys, have been carried out, and the general impression they leave is that the mine is in such a distant, remote area with low species diversity and no significant archaeological sites that the risk of impacts to the environment and wildlife would be minimal and easy to monitor. This runs counter to Ardener's paradox – extracting resources from the remote Citronen Fjord will, Ironbark assures, not expose the area to vulnerabilities that cannot be easily monitored and managed. In particular, the EIA largely mirrors the shipping assessment done in 1996, but it goes further by judging a shipping accident or marine fuel spill to be unlikely.

Should mineral extraction go ahead – construction and development of infrastructure and ancillary facilities will take two years – lead and zinc ore will be mined year-round and transported by ship through the pack ice of the High Arctic (and will pass near or through the Northeast Water Polynya) to Akureyri, Iceland, or another northern European port. An impact benefit agreement aims to ensure employment opportunities for Greenlanders in the project; people from Greenland employed at the mine will reach it via Pituffik Space Base while the plan is for foreign employees to be flown into the operations area from a location in northern Norway (for example, Longyearbyen in Svalbard is the closest civilian airport), likely via Station Nord, which remains a military outpost in Northeast Greenland, but has also become more recently established as a scientific research station. The airstrip at Station Nord has been crucial for exploratory operations since Platinova's first operations (Platinova, for instance, used Boeing 727s, Hercules, and HS 748 aircraft). As of the time of writing, mine site construction had yet to begin.

In calculating the mineral properties and potential of the Citronen Fjord resource space, geologists, consultants, and company employees draw upon older and more recent subterranean representations, analyse decades of aerial photographs and satellite images, and monitor the latest scientific research on melting sea ice. In this context, Bruno Latour's (1987) idea of cycles of accumulation seems apt when reflecting upon the processes by which geological strata and subsurface spaces are explored, measured, described, made legible, and controlled by those involved in projects such as the Ironbark venture. These processes disclose how northern Greenland's shifting topographies are increasingly apparent through diminishing glacial ice, the retreat and thinning of sea ice, thawing permafrost, and coastal erosion. They also assess how these conditions as well as an intensification of stormy weather and rough seas would enable or hinder extraction, mine operations, and shipping. The proponents of the Citronen Base Metal Project and their teams of expert consultants speak in positive terms about how melting – and disappearing – ice would allow for greater possibilities for shipping in and out of the fjord.

Ironbark's Bankable Feasibility Report from July 2021 states that, based on analysis of geophysical data from boreholes and other samples, the resource area is much larger than previously known, with indications of large mineral deposits deep down in as yet unrecognised layers of ore. As I discussed in Chapter 3, excited talk of northern places such as Citronen Fjord emerging as frontier resource spaces fills the offices and meeting rooms of civil servants, geologists, industry consultants, and entrepreneurs in Nuuk. Other areas of northern Greenland have been assessed as having significant mineral potential – for example, a lead-zinc-silver occurrence was discovered in Washington Land in 1997 (Jensen and Schønwandt 1998).

Those parts of northern Greenland that are imagined as remote outer edges of the country may be re-spatialised as accessible sites for development and production, but remoteness still matters as a way of framing and legitimising the marking out of resource zones. Geological assessments have often described Greenland's far north as a mostly barren Arctic desert (Jensen and Thorning ibid.). The site for the Ironbark mine lies within the Northeast Greenland National Park, which was established in 1974 and is the largest national park in the world; in company reports and impact assessments as well as in government discourse, much is made of the site's distance from the nearest Indigenous communities in Northwest and East Greenland – to say nothing of the vast spaces of the inland ice that separate it from the capital city.

The site very rarely figures in the conversations I have with friends, scientists, environmentalists, and activists, or even with politicians and business leaders in Nuuk, even though Ironbark and its consultants have organised public information sessions in the capital, and given the prominence of the project and its approval. This is not to say that some politicians are not excited about the project or that some members of civil action groups are not worried about it. Mostly though other extractive industry projects preoccupy them, such as the mines being scoped out in the Nuup Kangerlua region, oil exploration in Davis Strait and Baffin Bay, or the controversies over uranium mining and possible environmental rupture near Narsaq. Citronen Fjord is simply a place that is beyond the daily lives and geographical imaginaries of most people in Greenland – even local Nuuk-based NGOs and citizen action groups that are opposed to mining. It is out of reach to those other than the geologists, survey crews, scientific teams, and members of the Danish navy's Sirius Patrol – as well as expeditioners and mountaineers – who venture to what are considered to be the remote northern reaches of the country. This sense of remoteness serves to limit the visibility of the Ironbark project and pushes it beyond the edges of public interest. This makes it far easier to turn it into a sacrifice zone.

The Ironbark mine illustrates well how the making of abstract resource spaces involves practices and procedures of exploring and mapping subsurface formations as well as economic calculation that furnishes narratives about the successful operation of extractive industries in Greenland's 'far' north. All of this serves to render such spaces as wild zones that can be set aside for resource development projects with a negligible environmental and social impact. Meanwhile, further to the southwest, other extractive projects are being assessed and marked out as possible resource spaces. In Washington Land, for example, Ironbark is also exploring the prospects for a base metals project, while seismic surveys to assess possible hydrocarbon reserves have been carried out in Baffin Bay and Melville Bay in recent years in offshore waters close to several small communities that are dependent on hunting and fishing.

Resources and the spaces within which they are discovered, identified, measured, and dug out are assembled (Li 2014), socially constructed (Bridge 2009b), and calculated for their value. They are formed and abstracted by technocratic practices and political gestures that intimate a negligible impact on environment, animals, and society as a result of development in these zones, as illustrated by plans for extracting iron ore, lead, zinc, ilmenite, and rare earth elements. The contours of Greenlandic landscapes and seascapes and the geological strata formed and laid down in deep time underpin geospatial assemblages of resources, which have become a basis for contemporary ideas about sustainability that stretch far into an imagined future. International companies as well as Greenland's political authorities and business élites may see potential arising from deep within mountains and the seabed, but the possibility of extractive industries being developed makes many residents in communities that are dependent on hunting and fishing anxious. In the next chapter, I explore how this is so with reference to the coastal northwest.

Notes

1 See http://ironbark.gl/projects/greenland/citronen/. Accessed 26 January 2023.
2 See https://bluejaymining.com/projects/greenland/disco-nuussuaq/. The description of the project includes this statement about the area being assessed:

> Disko is circa the size of Luxembourg with the largest anomaly being over 6 km long. Bluejay intends to build upon its understanding of Disko and the specific geological targets in order to prove up the resource potential. Multiple occurrences of nickel and copper sulphide bearing boulders have been identified throughout licence holdings and its prospectivity has been highlighted by major mining companies recently acquiring c.10,000 km^2 of the licence area surrounding Disko.
> Accessed 17 January 2023

3 https://bluejaymining.com/projects/greenland/dundas-iimenite-project/. Accessed 17 January 2023.
4 See Bryn Gray, "Building Relationships and Advancing Reconciliation through Meaningful Consultation." Report to the Minister of Indigenous and Northern Affairs, 16 May 2016. Available at https://www.rcaanc-cirnac.gc.ca/eng/1498765671013/16 09421492929. Accessed 17 January 2023.
5 Mineral Resources Authority, https://govmin.gl/. Accessed 18 January 2023.
6 Some of the ideas in this section were first sketched out in Mark Nuttall (2022), "Wild lands, remote edges: formations and abstractions in Greenland's resource zones," in Arthur Mason (ed.) *Arctic Abstractive Industry*. Oxford: Berghahn, pp. 83–107.

References

Ackrén, Maria. 2016. "Public consultation processes in Greenland regarding the mining industry." *Arctic Review on Law and Politics* 17 (1): 3–19.

Adunbi, Omolade. 2020. "Extractive practices, oil corporations and contested spaces in Nigeria." *The Extractive Industries and Society* 7 (3): 804–811.

Andersen, Erik O. 2001. "Mining concept for the Citronen Fjord zinc deposit, NE Greenland." In Hans Christian Olsen, Lida Lorentzen and Ole Rendal (eds.) *Mining in the Arctic*. London: CRC Press, pp. 51–60.

Arboleda, Martín. 2020. *Planetary Mine: Territories of extraction under late capitalism.* London and New York: Verso.

Ardener, Edwin. 1987. "Remote areas: Some theoretical considerations." In Anthony Jackson (ed.) *Anthropology at Home*. London and New York: Tavistock, pp. 38–54.

Bainton, Nicholas and John R. Owen. 2019. "Zones of entanglement: Researching mining areas in Melanesia and beyond." *The Extractive Industries and Society* 6 (3): 767–774.

Bax, Vincent, Wendy Francesconi and Alexi Delgado. 2019. "Land-use conflicts between biodiversity conservation and extractive industries in the Peruvian Andes." *Journal of Environmental Management* 23 (2): 1028–1036.

Behrisch, Erika. 2003. "'Far as the eye can reach': Scientific Exploration and Explorers' Poetry in the Arctic, 1832–1852." *Victorian Poetry* 41 (1): 73–92.

Bergmann, Linda S. 1993. "Woman against a background of white: The representation of self and nature in women's arctic narratives." *American Studies* 34 (2): 53–68.

Boertmann, David. 1996. *Environmental Impacts of Shipping to and from Citronen Fjord: A Preliminary Assessment.* NERI Technical Report, no. 162. Roskilde: National Environmental Research Institute.

Bridge, Gavin. 2004. "Mapping the bonanza: Geographies of mining investment in an era of neoliberal reform." *The Professional Geographer* 56 (3): 406–421.

Bridge, Gavin. 2009a. "The hole world: Scales and spaces of extraction." *New Geographies* 2: 43–48.

Bridge, Gavin. 2009b. "Material worlds: Natural resources, resource geography, and the material economy." *Geography Compass* 3 (3): 1217–1244.

Chapin, Stuart, Mathew Berman, Terry V. Callaghan, Peter Convey, Anne-Sophie Crépin, Kjell Danell, Huch Ducklow, Bruce Forbes, Gary Kofinas, Anthony D. McGuire, Mark Nuttall, Ross Virgina, Oran R. Young, and Sergei A. Zimov. 2005. "Polar systems." In Rashid Hassan, Robert Scholes, and Neville Ash (eds.) *Millennium Ecosystem Assessment Ecosystems and Human Well-Being: Conditions and trends.* Washington DC: Island Press, pp. 717–743.

Cons, Jason and Michael Eilenberg. 2019. "Introduction: On the new politics of margins in Asia: Mapping frontier assemblages." In Jason Cons and Michael Eilenberg (eds.) *Frontier Assemblages: The emergent politics of resource frontiers in Asia.* Chichester: John Wiley and Sons Ltd, pp. 1–18.

Dahl, Parnuna P.E. and Anne Merrild Hansen. 2019. "Does indigenous knowledge occur in and influence impact assessment reports? Exploring consultation remarks in three cases of mining projects in Greenland." *Arctic Review* 10: 165–189.

Dawes, Peter Robert. 1991. "Lauge Koch: Pioneer geo-explorer of Greenland's Far North." *Earth Sciences History* 10 (2): 130–153.

Dodds, Klaus and Mark Nuttall. 2019 "Geo-assembling narratives of sustainability in Greenland." In Ulrik Pram Gad and Jeppe Strandsbjerg (eds.) *The Politics of Sustainability in the Arctic: Reconfiguring identity, space and time* London and New York: Routledge, pp. 224–241.

Espig, Martin and Kim de Rijke. 2018. "Energy, anthropology and ethnography: On the challenges of studying unconventional gas developments in Australia." *Energy Research and Social Science* 45: 214–223.

Fentiman, Alicia. 1996. "The anthropology of oil: The impact of the oil industry on a fishing community in the Niger Delta." *Social Justice* 23 (4): 87–99.

Fentiman, Alicia and Nenibarini Zabbey. 2015. "Environmental degradation and cultural erosion in Ogoniland: A case study of the oil spills in Bodo." *The Extractive Industries and Society* 2 (4): 615–624.

Ferguson, James. 2005. "Seeing like an oil company: Space, security, and global capital in neoliberal Africa." *American Anthropologist* 107 (3): 377–382.

Flemmer, Riccarda and Schilling-Vacaflor. 2016. "Unfulfilled promises of the consultation approach: The limits to effective indigenous participation in Bolivia's and Peru's extractive industries." *Third World Quarterly* 37 (1): 172–188.

Gerhardt, Hannes. 2018. "The divergent scalar strategies of the Greenlandic government and the Inuit Circumpolar Council." In Kristian Søby Kristensen and Jon Rahbek-Clemmensen (eds.) *Greenland and the International Relations of a Changing Arctic: Postcolonial paradiplomacy between high and low politics*. London and New York: Routledge, pp. 113–124.

Gómez-Barris, Macarena. 2017. *The Extractive Zone: Social ecologies and decolonial perspectives*. Durham, NC: Duke University Press.

Government of Greenland. 2016. *Social Impact Assessment (SIA): Guidelines on the process and preparation of the SIA report for mineral projects*. Nuuk: Ministry of Labour, Industry and Trade.

Grønnow, Bjarne and Jens Fog Jensen. 2003. *The Northernmost Ruins of the Globe: Eigil Knuth's archaeological investigations in Peary Land and adjacent areas of High Arctic Greenland*. Meddelelser on Grønland – Man and Society 29. Copenhagen: Danish Polar Center.

Günel, Gökçe. 2019. "Subsurface workings: How the underground becomes a frontier." In Jason Cons and Michael Eilenberg (eds.) *Frontier assemblages: The emergent politics of resource frontiers in Asia*. Oxford: Wiley, pp. 41–57.

Guzmán-Gallegos, María A. 2019. "Controlling abandoned oil installations: Ruination and ownership in northern Peruvian Amazonia." In Cecilia Vindal Ødegaard and Juan Javier Rivera Andía (eds.) *Indigenous Life Projects and Extractivism: Approaches to social inequality and difference*. Cham: Springer, pp. 53–73.

Hanrahan, Maura. 2017. "Enduring polar explorers' Arctic imaginaries and the promotion of neoliberalism and colonialism in modern Greenland." *Polar Geography* 40 (2): 102–120.

Hein, Carola. 2018. "'Old refineries rarely die': Port city refineries as key nodes in the global petroleumscape." *Canadian Journal of History* 53 (3): 450–479.

Herrmann, Thora Martina, Per Sandström, Karin Granqvist, Natalie D'Astous, Jonas Vannar, Hugo Asselin, Nadia Saganash, John Mameamskum, George Guanish, Jean-Baptise Loon and Rick Cuciurean. 2014. "Effects of mining on reindeer/ caribou populations and indigenous livelihoods: Community-based monitoring by Sami reindeer herders in Sweden and First Nations in Canada." *The Polar Journal* 4 (1): 28–51.

Hill, Jen. 2008. *White Horizon: The Arctic in the nineteenth-century British imagination*. Albany: State University of New York Press.

Hubbard, Rutherford 2013. "Risk, rights and responsibilities: navigating corporate responsibility and indigenous rights in Greenlandic extractive industry development." *Michigan State International Law Review* 22 (1): 101–166.

Jacka, Jerry. 2015. *Alchemy in the Rain Forest: Politics, ecology, and resilience in a New Guinea mining area.* Durham, NC: Duke University Press.

Jacka, Jerry K. 2018. "The anthropology of mining: The social and environmental impacts of resource extraction in the mineral age." *Annual Review of Anthropology* 47: 61–77.

Jensen, S.M. and H.K. Schønwandt. 1998. "A new carbonate-hosted Zn-Pb-Ag occurrence in Washington Land, western North Greenland." *Danmarks og Grønlands Geologiske Undersøgelse Rapport* 3.

Jensen, S.M. and L. Thorning. 2001. "Challenges to exploration in Greenland's High Arctic plains and plateaus." In Hans Christian Olsen, Lida Lorentzen and Ole Rendal (eds.) *Mining in the Arctic.* London: CRC Press, pp. 65–70.

Kapel, Hans. 1994. *Citronen Fjord, Nord Grønland: Arkæologisk recognescering i forbindelse råstofefterforskning i det nordøstlige Peary Land, sommeren 1994.* Report, Nuuk: Greenland National Museum and Archives.

Karlsson, Bengt G. 2011. *Unruly Hills: A political ecology of India's Northeast.* New York and Oxford: Berghahn Books.

Keeling, Arn and John Sandlos (eds.). 2015. *Mining and Communities in Canada: History, politics and memory.* Calgary: University of Calgary Press.

Kirsch, Stuart. 2014. *Mining Capitalism: The relationship between corporations and their critics.* Oakland: University of California Press.

Koch, Lauge. 1922. "Ethnographical observations from the southern coast of Washington Land." *American Anthropologist* 24 (4): 484–487.

Kragh, Karsten, Svend Monrad Jensen, and Henrik Fougt. 1997. "Ore geological studies of the Citronen Fjord Zinc Deposit, North Greenland: Project 'Resources of the Sedimentary Basins of North and East Greenland.'" *Geology of Greenland Survey Bulletin* 176: 44–49.

Kuletz, Valerie L. 1998. *The Tainted Desert: Environmental and social ruin in the American West.* New York and London: Routledge.

Labban, Mazen. 2014. "Deterritorializing extraction: Bioaccumulation and the planetary mine." *Annals of the Association of American Geographers* 104 (3): 560–576.

Langdahl, Bjarne R., and Bo Elberling. 1997. "The role of bacteria in degradation of exposed massive sulphides at Citronen Fjord, North Greenland: Project 'Resources of the Sedimentary Basins of North and East Greenland.'" *Geology of Greenland Survey Bulletin* 176: 39–43.

Larsen, Peter Bille. 2015. *Post-frontier Resource Governance: Indigenous rights, extraction and conservation in the Peruvian Amazon.* London: Palgrave Macmillan.

Latour, Bruno. 1987. *Science in Action: How to follow scientists and engineers through society.* Cambridge, MA: Harvard University Press.

Li, Fabiana. 2015. *Unearthing Conflict: Corporate mining, activism, and expertise in Peru.* Durham, NC: Duke University Press.

Li, Tania Murray. 2014. "What is land? Assembling a resource for global investment." *Transactions of the Institute for British Geographers* 39 (4): 589–602.

Mason, Arthur. 2012. "Industry as Alaska's ethnographic present." *The Polar Journal* 2 (1): 77–92.

Mason, Arthur. 2022. *Arctic Abstractive Industry: Assembling the vulnerable and valuable North.* Oxford and New York: Berghahn.

Nuttall, Mark. 2010. *Pipeline Dreams: People, environment and the Arctic energy frontier.* Copenhagen: IWGIA.

Nuttall, Mark. 2013. "Zero tolerance, uranium and Greenland's mining future." *The Polar Journal* 3 (2): 368–383.

Nuttall, Mark. 2016. "Narwhal hunters, seismic surveys and the Middle Ice: Monitoring environmental change in Greenland's Melville Bay." In Susan A. Crate and Mark Nuttall (eds.) *Anthropology and Climate Change: From actions to transformations*, London and New York: Routledge, pp. 354–372.

Pijpers, Robert. 2019. "Territories of contestation: Negotiating mining Concessions in Sierra Leone." In Robert Pijpers and Thomas Hylland Erikssen (eds.) *Mining Encounters: Extractive industries in an overheated world*. London: Pluto Press, pp. 78–96.

Pratt, Mary-Louise. 1992. *Imperial Eyes: Travel writing and transculturation*. London: Routledge.

Priebe, Janina. 2017. *Greenland's Future: Narratives of natural resource development in the 1900s until the 1960s*. Umeå: Umeå Universitet.

Rasmussen, Mattias Borg and Christian Lund. 2018. "Reconfiguring frontier spaces: The territorialization of resource control." *World Development* 101: 388–399.

Ries, Christopher. 2002. "Lauge Koch and the mapping of North East Greenland: tradition and modernity in Danish Arctic research." In Michael Bravo and Sverker Sörlin (eds.) *Narrating the Arctic: A cultural history of Nordic scientific practices*. Canton, MA: Science History Publications, pp. 199–234.

Schritt, Jannik and Andrea Behrends. 2018. "'Western' and 'Chinese' oil zones: Petro-infrastructures and the emergence of new trans-territorial spaces of order in Niger and Chad." In Ulf Engel, Marc Boeckler and Detlef Müller-Mahn (eds.) *Spatial Practices: Territory, border and infrastructure in Africa*. Boston: Brill, pp. 211–230.

Sorge, Ernst. 1935. *With 'Plane, Boat, and Camera in Greenland: An Account of the Universal Dr. Fanck Greenland Expedition*. London: Hurst & Blackett Ltd.

Tsing, Anna. 2005. *Friction: An ethnography of global connection*. Princeton and Oxford: Princeton University Press.

Van der Stijl, Frank W., and Greg Z. Mosher. 1998. "The Citronen Fjord Massive Sulphide Deposit, Peary Land, North Greenland: Discovery, Stratigraphy, Mineralization, and Structural Setting." *Geology of Greenland Survey Bulletin* 179. Copenhagen: Geological Survey of Denmark and Greenland.

Watts, Michael J. 2015. "Securing oil: Frontiers, risk, and spaces of accumulated insecurity." In *Subterranean Estates: Life worlds of oil and gas*, edited by Hannah Appel, Arthur Mason, and Michael Watts, 211–36. Ithaca, NY: Cornell University Press.

Whitehead, Alfred North. 2007. *The Concept of Nature*. New York: Cosimo. Originally published in 1920.

Willow, Anna J. 2012. *Strong Hearts, Native Lands: The cultural and political Landscape of Anishinaabe anti-clearcutting activism*. Albany: State University of New York Press.

Willow, Anna J. 2017. "Cultural cumulative effects: Communicating industrial extraction's true costs." *Anthropology Today* 33 (5): 21–26.

Yakovleva, Natalia. 2011. "Oil pipeline construction in Eastern Siberia: Implications for indigenous people." *Geoforum* 42 (6): 708–719.

Zimmermann, Erich W. 1951. *World Resources and Industries: A functional appraisal of the availability of agricultural and industrial resources*, 2nd edn. New York: Harper and Brothers.

6 Places of human and non-human encounters

Chapter 5 discussed – and reinforced my point about – how the planning processes for mining ventures and oil exploration in Greenland involve political, economic, and calculative practices of abstraction. These make, mark off, and allocate a particular form of corporate value to resource spaces in what are thought of and defined as remote areas at a considerable distance from human habitation, such as the Ironbark mine in Citronen Fjord. Subsurface geologies are assessed for their economic potential and classified as stocks and storehouses of ore and hydrocarbons. It is in these spaces where companies make plans for holes to be dug and from which things will be unearthed so that they can be turned into resources. In resource company-speak and in impact assessments, Greenland's extractive spaces are described as frontier zones. They are commonly depicted and represented in reports and public hearings as empty wilderness areas with no human presence or imprint and low in biodiversity, yet they are described as being filled with an abundance of subsurface resources that justify the considerable investment needed for lucrative mineral projects that promise to produce significant tonnage of commercial value.

In this chapter, I illustrate this further by drawing on my long-term research in Northwest Greenland's coastal communities, in the area that encompasses the Upernavik and Avanersuaq districts. I mentioned in the introductory chapter that this is a region in which I first carried out anthropological fieldwork in the late 1980s and where I continue to do ethnographic research. Since 2014, much of this research has focused on the effects of climate change – especially on sea ice, glaciers, and the coastal waters – and what this means for local livelihoods; but I have also been examining the interest expressed by extractive industry in the area, people's responses to it, and how they encounter the activities and traces of industry. People there often tell me they have been rarely consulted when it comes to discussions of resource development, environmental protection, and wildlife management, or the processes that lead to environmental and social impact assessments.

This is consistent with the situations I described in Chapter 5. For them, this is a disappointment and they feel that meaningful consultation would involve a conversation with government and industry about the accepted technocentric and scientifically informed ways of viewing the world and would also allow for

DOI: 10.4324/9781003175421-7

an appreciation of the ways people live within surroundings they share and co-constitute with the non-human (cf. Kohn 2013). Indeed, they say that Indigenous perspectives on non-human entities – and how the non-human speaks to us about being in the world – should also be considered in participatory processes and need to be taken into account by consultants who conduct social and environmental impact assessments. These local concerns highlight the temporal and spatial lim-itations of social and environmental impact assessments and point to the need for dialogue on how Indigenous ontologies can contribute to "a more expansive, inclusive approach" (cf. Behn and Bakker 2019: 99).

The northwest coast

Northwest Greenland comprises the two former municipalities of Upernavik and Qaanaaq (the latter is also known as the Avanersuaq area), which are part of Greenland's largest municipal region of Avannaata Kommunia. The administra-tive headquarters are located at Ilulissat in Disko Bay in Central West Greenland and the municipality stretches northwards (and includes vast parts of the inland ice sheet) to Nares Strait, which connects northern Baffin Bay with the Lincoln Sea in the Arctic Ocean. There was a reorganisation of Greenland's municipal-ities in early 2009, which meant local and regional decision-making was moved away from the towns of Upernavik and Qaanaaq to Ilulissat. Despite this, both areas retain identities as distinct districts of northern Greenland and livelihoods remain, for the most part, based on hunting and fishing. Indeed, there are specific marine mammal hunting and fisheries management regimes and quota systems that apply to these areas, and so I refer to them by their old district names as people living there continue to do.

While people identify as Kalaallit and Inuit, the population in Avanersuaq are known and refer to themselves as Inughuit. Many people in Upernavik and Avanersuaq often say they feel they also live in regions that are far removed from political concern and daily decision-making in Ilulissat and even further from Nuuk. They talk about how they think Northwest Greenland and people's daily lives and their interests appear peripheral and distant to politicians and decision-makers in these more southerly, increasingly urbanised centres of the country (Nuttall 2017). They imagine that seen from Nuuk, Upernavik, Qaanaaq, and the other smaller communities of the northwest must be thought of as remote, far off places lying at the very outer edge (*nunap isua*) of the country.

The Upernavik area is a 450 kilometres-long stretch of coast. It extends from near the northern coastline of Sigguup Nunaa (the Svartenhuk peninsula) in the south to Qimusseriarsuaq (Melville Bay) in the northern part of Baffin Bay. About 2,800 people live in the Upernavik district – the town of Upernavik has a population of around 1,100 and some 1,700 people inhabit nine smaller villages, ranging in size from about 50 in Naajaat to 450 in Kullorsuaq (Greenland's largest village). In the Avanersuaq area, the population is around 800. Qaanaaq's pop-ulation is about 650, while some 60 people live in Savissivik, around the same number live in Siorapaluk, and some 25 in Qeqertat. There are strong networks of kinship and intricate forms of social relatedness throughout both the Upernavik and Qaanaaq regions, with the communities of Savissivik and Kullorsuaq having especially close family connections. There is regular movement and the exchange and sharing of hunting and fishing products between them, with people crossing the sea ice in Melville Bay by dog sledge in winter and spring and travelling on the sea by open boat during the summer. Upernavik and Qaanaaq are supply centres for the villages in each district and a number of public sector services and private businesses provide some employment.

For anthropologists, Avanersuaq is a much-visited site of ethnographic enquiry (Hastrup 2019), but it has also long been viewed as a scientific field site. The Danish Meteorological Institute (DMI), for instance, maintains a geophysical observatory in Qaanaaq which was established in the 1950s. Early work there focused on geomagnetism, which is ongoing today, but in recent years DMI has developed a participatory ocean and cryosphere monitoring programme in collaboration with local hunters. Qaanaaq has also become an especially popular choice for television documentary makers, journalists, and researchers who are eager to film and write about the effects of climate change and chronicle what they see as a small group of traditional Inuit hunters clinging to an ancient way of life in the modern world, and on diminishing and increasingly thinner ice. Scientists from around the world visit there each summer to study the area's glaciers, the inland ice, and fjord systems; social scientists study community vulnerability and resilience; and tourists arrive each spring to be taken out on trips by dog sledge on the sea ice by local hunters. Climate change has given this interest and attention an added urgency and the region has come under greater scrutiny as a global hotspot. It has assumed significance as a place for observation and monitoring and is considered a barometer of global change.

The Upernavik district also draws scientists to study the inland ice there and its outlet glaciers, particularly the Upernavik Isstrøm, which is a set of major tidewater glaciers that have become especially important to observational studies of ice thickness, glacier discharge, and terminus retreat. Part of this importance also lies in a rich historical record of terminus positions of the Upernavik Isstrøm that dates from 1849 (Haubner et al. 2018; Weidick 1958). Similarly represented as remote and traditional, although not as alluring for journalists, film crews, and researchers as much as Avanersuaq, where it is assumed hunters have clung to more 'traditional' hunting methods and ways of life, Upernavik has also seen some limited tourism in recent years (mainly in the form of cruise ships, some of which continue north across Melville Bay to Qaanaaq and Siorapaluk, which is the most northerly settlement in Greenland). However, a lack of accommodation and the erratic nature of the air link to Ilulissat, which is often affected by poor weather, mean that it is still a part of Greenland that can be difficult (and expensive) to reach and travel around.

Despite its attraction as a region where climate change can be observed and the impacts of globalisation on a small hunting community can be studied, the old Thule district has long been attractive to explorers, adventurers, and scientists. Still constructed as a remote, exotic place in the global imaginary, it is now viewed as one of the last places in the Arctic to experience what are often described in popular accounts as the last days of traditional Inuit hunting culture, even if Jean Malaurie (1982) wrote about it as such in *The Last Kings of Thule*, the narrative that resulted from his sojourn among Inughuit in the early 1950s (he was also witness to the construction of Thule Air Base), or as others such as Marie Herbert (1973) depicted the region in *The Snow People*. As Hastrup (ibid.: 51) puts it, Inughuit are "already figuring as lost and turned into heritage for protection." But examining the history of such accounts – for they have a value in understanding the construction and representation of the district – it would appear that Inughuit have long been written about as always living in a precarious time, their culture and livelihood as hunters threatened with disappearance.

I once wrote the text to accompany a book of photographs by celebrated Icelandic photographer Ragnar Axelsson (2010), who is known widely as RAX. His mainly black and white images were taken during many trips he made over several years to Avanersuaq and East Greenland, and they were gathered together in a book titled *Last Days of the Arctic*. They are beautiful, evocative, and often stark. I had no say in the title and the original idea by the Icelandic publisher was for me to write text that was much longer than appeared in the final version of the book. And it was this longer text that I generated. However, once I had done this, they told me that European co-publishers wanted to reduce the number of words in the text so that the focus was much more on the photographs and what RAX intended them to represent – exactly what the book's title suggests, that we may be witnessing the last days of the Arctic as global forces and environmental catastrophe collide in the region. It was felt that this would appeal to a wider market with a readership eager to read about a place under threat from climate change and about a people who were disappearing because of it. The result was that my text, which, while drawing attention to the history and legacy of change, was shortened considerably. I wanted readers to be aware of what Inughuit and other Inuit groups had experienced, but I did not situate my words within the scholarly and more popular traditions of writing about the area – and the wider Arctic – as lamentation. It is far too easy to write about the Arctic and its peoples in a way that contributes to and reinforces a categorisation of it as a disappearing world. I felt frustrated however that I may not have got across the point in the way I had intended to do – that we must understand Indigenous resilience and adaptive responses, as well as Indigenous agency, and that Inuit are not mere victims of change who are living in vulnerable circumstances.

The mixed economies of the smaller settlements (and many household economies in the towns of Upernavik and Qaanaaq) are based largely on the procurement of living marine resources – marine mammals such as seals, walrus, polar bears, narwhal, beluga, fin and minke whales, and fish such as Greenland halibut, cod, salmon, Arctic char, Atlantic wolfish, and capelin. People also depend on land animals such as musk oxen and reindeer – and on some full-time, part-time, or seasonal work. *Kalaalimernit* (Greenlandic foods) enter and sustain an informal economy. Many of the catch shares – from seals, walrus, and Greenland halibut, for instance – circulate within and around families, households, and communities, but people sell some of the things they hunt and fish. Meat and fish products also find their way into and around local distribution channels and provide the basis for a formal economy (alongside the informal one of procurement, sharing, and reciprocity that is characterised by kinship and close social association) which gives people the opportunity to earn some of the money necessary for maintaining a hunting and fishing way of life. And it is the inshore fishery for Greenland halibut, carried out by using longlines from small boats in summer and snowmobiles on the ice during winter and spring, that is the most important commercial fish stock for the mixed economy.

The halibut catch provides the cash that is needed for the materials and equipment that make hunting and fishing possible in the first place – rifles, bullets, nets, fishing hooks, outboard engines, fuel, snowmobiles, clothing and so on – as well as for covering the cost of daily living (Delaney et al. 2012; Nuttall 1992). The Greenland halibut fishery in Upernavik is regulated by a management system that quite often disadvantages local hunters and fishers who use small boats. Large fishing boats are assigned quota shares which are transferable – fishers can sell their share or quota size or buy them from other fishers. Many of the large fishing vessels that operate in the waters of the Upernavik district come from further south, and so are often in competition with the smaller boats that focus on the inshore fishery. A number of large fishing boats sail north from Disko Bay, where there are concerns that the fishery has overexploited the Greenland halibut population (e.g. Fredenslund 2022).

Figure 6.1 Kullorsuaq, Upernavik district (Photograph Mark Nuttall).

People live, work, and move around a complex topography of islands, headlands, fjords, sea, tidewater glaciers, the edges of the inland ice, mountains and fells, lakes and rivers, and in winter and spring, sea ice. A configuration of multispecies encounters and engagement with the non-human characterises both the human history of the region and contemporary human-environment relations. Humans, animals, fish, snow, ice, water, rocks, wind, storms, wave action, and everything else that makes up and moves within and through these Arctic surroundings are active participants that share and shape them. Local knowledge of the multi-layered and textured places in which hunting and fishing activities occur and through and around which people move and travel is extensive, rich, and deep. Human-environment relations arise from, are given meaning, and are reproduced through the interweavings, entanglements, and trajectories of human and non-human entities and the rhythms and flows, as well as the often difficult challenges and uneven turns, of everyday life on water, ice, and land.

The Upernavik district is described geographically as an archipelago, largely because it is characterised by numerous small islands, while Avanersuaq can be described more as a stretch of coast with a deep fjord, as well as a number of islands. I have often thought that Philip Hayward's (2012) idea of the 'aquapelago,' which he calls an assemblage of marine and terrestrial spaces, seems an apt and inspiring description for Northwest Greenland, or indeed any part of Greenland's coast for that matter. For Hayward, aquapelagic assemblages "are constituted by social units in locations where the aquatic spaces between and encircling islands are fundamentally interconnected with and essential to communities' inhabitation of their locale (and substantially generate their senses of identity and belonging to that place)" (Hayward 2015: 84). The aquapelago as assemblage, he argues, involves the interactions and entanglements between humans and other actants that "may be animate (living) entities, inanimate ones (such as sand, soil, etc.) or the product of energies (such as individual weather events or larger climatic patterns, such as global warming)" (ibid.). Northwest Greenland is a cryospheric world in which the ocean depths, sea ice, glaciers, and icebergs, the coastal interior landscapes of mountains and deep valleys, the subterranean and the seabed, and the atmospheric intermingle with the lives of humans and animals.

But the Northwest Greenland aquapelago is just one assemblage within a dynamic, unfolding, and emerging world which people living there think and talk about as well as relate to as *pinngortitaq*. Introducing this in Chapter 2, when I discussed the 'Inuit Pinngortitarlu' project in Nuup Kangerlua, I mentioned that *pinngortitaq* is often just translated from Greenlandic as 'environment' or 'nature.' However, it has a meaning that refers to all that is underneath, above, and all around. It refers to more than just the surface of the earth – it encompasses water, ice, soil, rock, sky, and wind; surroundings which include the air, atmosphere, subsoil, mountain interiors, and earth processes; what is above and below and around, and how all of these things intersect, interact, and are entangled. *Pinngortitaq* is always coming in being, forming, and reforming. To live and move within and across *pinngortitaq* is to have a continual encounter with its many moods and habits. One has to come to terms and engage with its dynamic, flourishing nature.

Climate change is evident in Northwest Greenland in quite often visually dramatic ways. Scientists, of course, study and write about this extensively, and the people living in the region experience it in more immediate ways than the analysis of quantitative data and satellite images can tell us. Given that my experience in the region now dates back a few decades, I would also say that to think of these transformations as dramatic is not an overstatement. The overall extent of sea ice cover in winter and spring is diminishing (AMAP 2012; Stroeve et al. 2017), coastal glaciers are melting and retreating, and iceberg calving rates are on the increase (Cowton et al. 2018; van As 2011). Observation and monitoring of Greenland's inland ice reveals that it is experiencing more surface melt and mass balance loss (van den Broeke et al. 2017). Briner et al. (2013) describe how the increased meltwater run off from glacial fronts has been affecting water temperature and circulation patterns as well as the formation of sea ice for several years. This has a profound influence on the distribution, movement, and availability of the marine mammals and the fish people depend upon as the basis for local livelihoods.

In summer and autumn, some seals have been observed by hunters in the Upernavik district to move further away from coastal waters with the shifting pack, while changing ice conditions and warming waters also mean differences in the migration routes of seals, other marine mammals, and fish such as halibut and cod (Nuttall 2019a). As Laidre et al. (2016) show, the use of glacial fronts by narwhals in Melville Bay has expanded over the last two decades, probably because of reduced summer fast ice and a delay in autumn freeze-up. The study's results reveal that narwhals prefer glacial fronts with a higher freshwater melt rather than silt-laden discharge. Marine-terminating glaciers play a vital role in the productivity of Greenland's fjord systems. They provide nutrient-rich habitat for marine mammals, fish, and seabirds – and so glacial fjords become increasingly vital for ice-dependent species as sea ice disappears. Glacial fjord systems are also vital for sustaining important inshore and coastal fisheries, such as the Greenland halibut fishery, but the effects of glacial retreat to shallow water and an eventual ecosystem switch to them being land-terminating glaciers will likely have significant socio-economic implications (Laidre et al. 2015; Meire et al. 2017). As climate change intensifies, the scientific view is that eventually a number of glaciers will terminate on land rather than at the coast, and so will no longer calve icebergs which provide the source of freshwater for many communities in Northwest Greenland.

Areas of open water are widening in late winter and spring and this means that spending time at a shifting ice floe-edge (*sinaaq*, where the boundary between ice and open sea is less distinct) brings increasingly fewer returns for hunters. Kangerlussuaq (Inglefield Bredning), for example, is the main fjord system in the Qaanaaq area. It opens into the North Water Polynya/Pikialasorsuaq and northern Baffin Bay (I will return to discuss this ecosystem in Chapter 7). The changing nature of *sinaaq*, the ice edge, means it is not only a far more difficult and uncertain environment for hunting, but travelling to it from Qaanaaq in late winter and during spring is increasingly dangerous as the sea ice is not only thinner, headland cracks appear with increasing frequency because of changing glacier-ocean interactions in the fjord and present obstacles to movement across it by dog sledge in the form of wide open leads. Not much money can be earned as a hunter in the Avanersuaq and Upernavik districts – low prices are paid for sealskins by Royal Greenland, and the international trade of other marine mammal products, such as narwhal tusks and walrus ivory, is prohibited by CITES (the Convention on International Trade in Endangered Species of Wild Fauna and Flora). Combined with the increasingly erratic sea ice conditions that make ice edge hunting harder, this has meant harsh economic conditions for many households in Qaanaaq and the other villages of the district, which are now often in tighter financial circumstances.

Because the marine ecosystem supports and sustains local livelihoods, the changing nature of sea ice – how it forms, takes shape, and persists over winter and through spring – inevitably has consequences for hunting and fishing activities and local economies as well as for community life. Greenland halibut – a vital resource for earning money in the Upernavik district, as I explained above – are moving further north, for instance, while hunters report that seals are becoming scarce in some community waters. While this may have implications for the fishery in Upernavik, opportunities have arisen for the development of a Greenland halibut catch in Kangerlussuaq precisely because of those fish movements farther north. An experimental fishery carried out with longlines through holes in the ice has been carried out for the last nine or ten years during winter and spring and represents a good example of how people in the region can respond proactively to climate change (e.g. Andersen and Flora 2019).

At the time when the fishery was first being tested, hunting families began to talk about the environmental changes they were witness to in their surroundings as well as the effects of social and change they were experiencing in their communities. They discussed options for the future. Women in particular expressed concerns that as Qaanaaq's environment, economy, and households change and as gender roles are transformed, younger women were no longer learning skills necessary to support their male partners who were hunters; and they were especially anxious that the hunting way of life would erode further. A community partnership between the Qaanaaq hunters and fishers' association and Royal Greenland, the government-owned fishing company, established Inughuit Seafood in October 2014. This incorporated the community's existing storage freezer into Royal Greenland's seafood distribution system, providing opportunities for Qaanaaq to export its products elsewhere in Greenland and beyond. The construction of a fish freezing facility has made it possible for hunters to fish and land, and sell, their catch locally in Qaanaaq. Around 60 hunters/fishers from Qaanaaq and other settlements now travel to places in the Kangerlussuaq fjord system by dog sledge for several days or even a week or two at a time during winter and spring, spending time at fishing camps in small cabins placed on sledges or in tents; they travel back to Qaanaaq to sell Greenland halibut to the facility. With this new market, fewer hunters now travel out to *sinaaq* to hunt marine mammals, preferring instead to invest their time and energy in what to them are more profitable fishing activities (see also Flora et al. 2018).

Climate change, however, has to be understood in a wider context of other drivers of social and economic change, as well as politics and governance that affect daily life in Upernavik and Qaanaaq. For instance, management regulations and quotas for living marine resources combine with environmental and climate change to make for circumstances that restrict the abilities of local people to travel and move around their localities and to hunt and fish certain species. As the Greenland halibut fishery in the Qaanaaq area is still an experimental one, it is not yet covered by the rules and regulations of stock assessment and national fisheries management, as it is in other parts of Greenland, including the Upernavik district. Getting halibut from Qaanaaq to reach domestic and overseas markets is critical for the development of the fishery, but this also introduces greater elements of risk and vulnerability and perhaps makes people and their livelihoods less resilient to coping with and adapting to climate impacts. The Greenland halibut fishery may hold the promise of higher incomes for some hunters in Qaanaaq, but the catch must remain stored and frozen until a ship can travel north to the town in summer. As such, there is limited capacity in the freezing facility. Furthermore, exploratory activity related to extractive industries in Northwest Greenland – which people worry affect marine mammals and fish – has brought different kinds of pressures and anxieties as well as hopes for the future.

Sea ice extent may be declining rapidly in Northwest Greenland, but a point that needs to be emphasised is that having to contend with environmental and social change is nothing new for the region and its residents. Seasonal variations and fluctuations in sea ice have always posed challenges to people and their livelihoods. Furthermore, the legacies of colonialism, the impacts of commercial whaling by Europeans on northern marine ecosystems, the KGH trade system, the resettlement of people by the Danish state, and economic transitions, or international environmentalist opposition to marine mammal hunting, for example, continue to have their effects. However, the scale and extent of the changes people are witness to and the more extreme weather conditions they now experience on a daily basis, and how these conditions affect sea ice and coastal waters and influence the behaviour and habits of animals, bring a new set of challenges to life in northern places.

Mapping and surveying

In Northwest Greenland, having good access to hunting marine mammals from sea ice or by marine craft during the open water season has been a prerequisite for the survival of historic and contemporary communities. The rich and productive Arctic ecosystems, together with a considerable degree of mobility and flexibility in settlement patterns, resource use practices, and in the ways people organise their social lives, have enabled hunting societies to live and thrive from coastal resources since the first Paleo-Inuit and Thule culture migrations (Hansen 2008; Hastrup et al. 2018; Nuttall 1992; Petersen 2003). The region has long been viewed as a resource space by Indigenous hunters and more recent arrivals, including explorers, whalers, traders, geologists, and the mining and oil industries (Hastrup 2019; Nuttall 2017; Vaughan 1991). A point to which I will return in Chapter 7, however, is that the marine mammals, fish, and land animals that people living in Northwest Greenland rely on are not viewed merely as economic resources. 'Wildlife' is not an Indigenous categorisation that informs local understandings of human-environmental relations and multispecies encounters.

The North Water Polynya/Pikialasorsuaq (again, see more on this critical ecosystem in Chapter 7) and the area around Nares Strait constituted a gateway for the first Paleo-Inuit groups to cross into Greenland (Grønnow and Sørensen 2006; Hastrup et al. 2018). The Norse also sailed there and beyond from their settlements in Southwest Greenland as far north as the east coast of Canada's Ellesmere Island to hunt walrus. A runestone was found in the nineteenth century on the island of Kingittorsuaq to the north of the present-day site of Upernavik town, and there is possible archaeological evidence of a meeting and perhaps trade between Norse and Inuit in the Upernavik district. Hansen (2008) suggests that such contact was probably short-lived and that hunting families in the area were geographically isolated and did not encounter much disturbance from outside until whalers began to arrive.

Dutch and English whalers sailed frequently to Greenland's seas in the late seventeenth century, expanding their activities to central and northern waters, while English and Scottish whalers were regular summer visitors to Upernavik and later to Avanersuaq, from the nineteenth century until the first decade of the twentieth, venturing there as they headed further north to Melville Bay and the whaling grounds of Canada's Lancaster Sound in search of the Greenland right whale. Overexploitation of these northern waters led to the scarcity of the right whale, and by the first decade of the twentieth century in attempts to make their voyages profitable, whaling ships were returning from Greenland and Arctic Canada to ports like Dundee with the oil and bones from narwhals, beluga, walrus, and seals, and with polar bears.

Danish trade expanded into the Upernavik area from the late eighteenth century with the activities of the KGH. The settlement of Upernavik was established as a trading station in 1772, and Inuit hunting families visited there regularly during their seasonal round to trade blubber, sealskins, narwhal and walrus tusks, fox furs, and whalebone. Some moved there permanently. When I lived in the Upernavik district in the late 1980s, I knew people in settlements such as Kangersuatsiaq, Nutaarmiut, Nuussuaq, and Kullorsuaq who were then in their 60s, 70s, and 80s who told me stories they had heard first-hand from their parents and grandparents about the annual visits of British whalers. From the accounts I heard, their arrival in spring was something people looked forward to. At that time of year, Inuit would hunt narwhal, seals, and walrus, and would travel to the outer islands by umiaq and kayak to trade with the whalers. I heard stories about whalers staying in people's homes because of shipwreck, and people recalled that there were a number of their ancestors who learned English during their encounters with these seasonal visitors (see also Hansen ibid.: 22). However, the trade that went on was not always welcomed by the KGH station managers in the district as the whalers often had superior goods, especially in the form of kettles, pots, pottery ware, and tools, than the posts at the Danish colonies and *udsteder* were provisioned with.

In the Avanersuaq region, European and American explorers became regular visitors throughout the nineteenth century and early twentieth century during voyages to find the Northwest Passage and on expeditions to discover a route to the Arctic Ocean and the North Pole. Between 1817 and 1820, Scottish and English whalers began to sail through the pack ice of Melville Bay to the North Water. They crossed over to Lancaster Sound and the east coast of Baffin Island where they exploited new whaling grounds. Whaling fleets continued to reach the North Water throughout the nineteenth century and often stopped along the Avanersuaq coast.

Despite the activities of whalers and the presence of explorers, including Robert Peary who used the area as a base for his expeditions between 1891 and 1909 to explore northern Greenland and attempt to reach the North Pole, the district was not an area that interested the Danes as a potentially lucrative trading region. The KGH had established trading stations in the northern part of Upernavik district, but because of the smaller and scattered nomadic population of Inughuit, as well as the difficulties of navigating through sea ice, it did not think the areas up to Melville Bay and beyond held much promise – or were worth the investment needed to maintain colonies. This left ethnographer-explorer Knud Rasmussen a degree of freedom to establish an independent trading post as a commercial enterprise at Thule in 1910. Rasmussen used the profits from this regional trade to support administrative costs and provide some social services to the Inughuit, but also to finance many of his journeys in northern Greenland and Arctic Canada, which became known as the Thule expeditions (Vaughan ibid.). Danish colonisation may not have affected Avanersuaq directly during this period, but the presence of explorers, whalers, and independent traders meant the region "opened up to new forms of local exchange and long-distance trade that brought the Northwest Greenlanders into the global market economy" (Hastrup 2019: 43).

In Chapter 4, I referred to how, in the middle of the twentieth century and during the early years of the Cold War, the Inughuit of Uummannaq experienced upheaval and dispossession when they were moved and resettled in the new community of Qaanaaq in 1953 because of the expansion of US activities and infrastructure at Thule Air Base. The Upernavik and Avanersuaq districts were not affected to the same extent by the G-50- and G-60-initiated actions of the development of a commercial fishing industry and the closure of settlements as were areas further south. However, at one point, Danish administrators had entertained relocating the entire population of the Upernavik area to larger towns on the west coast. However, between the 1950s and 1970s, as elsewhere in Greenland, social change did follow from Danish modernisation policies, and since 1979, Northwest Greenland has been affected by transformations in governance resulting from Greenlandic home rule and self-government policy.

In both the Upernavik and Avanersuaq districts, many settlements were abandoned in the twentieth century by their inhabitants or closed by the former KGH because they no longer offered good hunting prospects or constituted viable trading places (Hansen ibid.; Petersen ibid.). Historically as well as in the present, living costs have been subsidised by the Danish authorities and more recently by the Greenlandic government, although subsidies for such necessities as heating oil and fuel are gradually being removed. In the 1980s, European- and North American-based animal-rights organisations mobilised resources to campaign effectively against seal hunting and the sale of sealskins. The legacy of this activism persists forty years later; in Northwest Greenland, the loss of markets for sealskins has dealt a significant blow to local economies, but the animal-rights and anti-sealing campaigns continue to be spoken about as an assault on Inuit society and culture.

The people of Northwest Greenland have thus been witness to and have been affected by a range of social and economic changes, as well as political action and decision-making, some of which have been profoundly disruptive. The region has been a meeting place for different cultures for several thousand years, beginning with the Paleo-Inuit and Thule movements and on to the visits of explorers, whalers, Danish administrators, and the US military. We can reflect on these historical and contemporary events, exchanges, interventions, and intrusions as encounters and antagonisms between different kinds of assemblages, in the sense that Deleuze and Parnet (2007) see an assemblage as a multiplicity of diverse but constituent parts. For Deleuze and Guattari (1988), these components are entangled with, and are affected by a complexity of processes that stabilise or destabilise them – and they can be disturbed further by components of other assemblages that collide, clash and even intermingle with them. For example, explorers and whalers brought their own ideas about the Arctic and marine mammals as well as different technologies of navigation, cartography, and resource exploitation, as well as trade items, into Northwest Greenland. They encountered and interacted with Inuit who had their own rich knowledge about and relations with place, animals, and the non-human world (Dodds and Nuttall 2019).

When it did eventually happen, the longer reach of the Danish administration into the northwest – especially in the Avanersuaq region – was enabled by the expansion of trade activities and the establishment of settlements as far as the southern part of Melville Bay, but also by geological mapping and cartographic surveys. These were vital for acquiring knowledge of Greenland's territorial extent, for determining its insularity, for the exercise of political power over the furthest reaches of the kingdom, and for laying the foundations for the later assessment of resource potential (cf. Akerman 2009; Edney 2009). In the first half of the twentieth century, geology was brought into the service of the state to assist in a more through exploration of Greenland's less charted areas and support assertions of Danish sovereignty. As I discussed in Chapter 1, given Norwegian interest in Northeast Greenland, along with the activities of trappers and hunters, the Danes could no longer afford to ignore the more northerly stretches of the country, including the northwest, as a region that was a constituent part of the kingdom.

Matthew Edney (ibid.) reminds us that maps serve as technologies of governance, with governments using them as devices to transform land into state territory. Robert Stafford (1984) points out that as the geological map developed as the centrepiece of geology's media, it assumed importance for showing the positions of rock masses and ore bodies, and gave "scientific order to the earth's surface" (ibid.: 5). Mapping sought to visualise and explain complex processes occurring over millions of years. But geology also acquired a territorial dimension from the nineteenth century, and through mapping, it expanded into and beyond the imperial sphere. Geology also became closely associated with military activities of terrain investigation and evaluation (Rose, Ehlen and Lawrence 2019), and emerged as a science concerned with territoriality, a process by which geological exploration helped to organise and order the vertical and volumetric dimensions of the state (Klinger 2015). But this has not been an exclusively Euro-American or as some might put it, a Western scientific imperial approach to territorial acquisition or the appropriation and commodification of the subsurface. Grace Yen Shen (2014), for instance, traces the development of geology in early twentieth-century China and shows how the country's pioneering geologists, in collecting rocks, fossils, and studying earth processes, served broader processes of nationalism. Geology, she argues, was essential to the constitution of land as "territory, resource, physical environment, and native place" and "science and nation converged in geological activity" (ibid.: 4).

Mapping – including the production of geological maps – became a national project that was important for Danish claims of possession, ownership, uccupancy, and sovereignty over North and East Greenland. Knowledge of northern Greenland's geological structure and mineral resources was vital for Danish consolidation of its Arctic territory, and so geologists and other scientists went about their research as part of a process of geo-securing Greenland's remoter and more unknown regions. The expeditions of geologist Lauge Koch, especially in the 1920s and 1930s, and which I referred to in Chapters 1 and 5, were critical to this endeavour (Ries 2002). And, as Chapter 4 discussed, discovering what lay under the ice, the land, and the seabed also became central to much scientific research carried out in Greenland by the US during the Cold War, and the northern parts of Upernavik district, especially around Melville Bay, and the Avanersuaq district were under sharp focus. Denmark also continued its own geological survey work, as did mining and oil companies, from the 1960s. Today, geological research is not only vital for the mapping and assessment of Greenland's resource potential, it has a geopolitical dimension in terms of the delimitation of the continental shelf and the nature of Greenland as a state in formation. As Bobbette and Donovan (2019: 2) put it, "geologists are themselves politicians operating in spaces" they seek to secure on behalf of others. Geology, they say, "emerges in and through political processes" (ibid.: 3).

For Bruce Braun (2000), the 'geologizing' of an area is often paired with 'economic and political rationality' and this is evident in how, again as I discussed in Chapter 1, Koch's geological cartographic work in Northeast Greenland was also vital for Denmark's case against Norway's assertions at The Hague in 1933, while also helping to establish certainty about the insularity of Greenland. Along with earlier expeditions to Northwest Greenland led by Knud Rasmussen and Northeast Greenland, such as the fated Danmark Expedition of 1906–1908, led by Ludvig Mylius-Erichsen (who died along with the Greenlandic catechist and explorer Jørgen Brønlund and Danish explorer and cartographer Niels Peter Høeg

Hagen), Koch's long periods of fieldwork were used by Denmark to justify a claim to sovereignty that rested not just on cartographic knowledge, but of Danish presence through scientific activity in what were thought of as Greenland's otherwise uninhabited, remote places. Koch laid the significant groundwork for determining the main geological elements of northern Greenland, but he and others also imprinted Danish expeditionary lines on the maps that were produced following their travels and surveys. Sledge routes and journeys by boat, campsites, shelters, and huts marked the movements and discoveries of Danish explorers and along with early archaeological assessments, helped in the assertion of claims that the region was a dwelt-in space regularly criss-crossed and traversed by Danes and Inuit – not the terra nullius the Norwegians said it was. In this way, and along with the later use of aircraft and the construction of scientific field stations and airstrips from the 1930s, Greenland's High Arctic spaces were filled in with evidence of historic and contemporary human activity and presence and became thought of as an integrated part of the Danish Realm (Ries ibid.).

Northwest Greenland as extractive space

Mineral exploration to assess the potential value of northern Greenland for resource extraction *per se* was of minor interest when the KGH was expanding its activities into the Upernavik district in the nineteenth and twentieth centuries. Some graphite had been mined at Akia (*Langø* in Danish), an island near Upernavik in 1845 (Ball 1923), while samples of sphalerite were collected on Red Head in Melville Bay at the beginning of the twentieth century (Bøggild 1929). In more recent decades, the region has been subject to a number of geological investigations into mineral occurrences and hydrocarbon potential. Oil exploration took place in northern Baffin Bay in the 1970s and again in the early 1990s, while four short reconnaissance trips by mining companies were made in Upernavik between 1969 and 1981 and mapping work was done by the former Geological Survey of Greenland (GGU) in the late 1970s. However, geologically, the Upernavik district remained one of the least explored parts of Greenland until summer 1998, when an extensive geological reconnaissance was carried out. A joint project between the Geological Survey of Denmark and Greenland (GEUS) and the BMP, it was funded by the Government of Greenland with the purpose of attracting the interest of the mining industry (Thomassen et al. 1999). Further north in Avanersuaq, GGU carried out an extensive mapping programme in the Peary Land region between 1978 and 1980, and followed this in 1984 and 1985 with a systematic survey of the area between J.P. Koch's Fjord, which is on the western coast of Peary Land, to Washington Land, north of Qaanaaq (Henriksen 1985). For a point of clarity on the historical organisation of geological survey work, GGU was established in 1946, the Geological Survey of Denmark can trace its origins to 1888, and the two were merged to form GEUS in 1995.

Fieldwork was carried out by GEUS in Inglefield Land in 1995. This followed an airborne magnetic and electromagnetic survey in 1994, which led to considerable economic interest in the area. This work indicated the presence of a large sulphide mineralisation and spurred on further surveys that investigated the potential for kimberlite pipes, including commercial exploration by RTZ in 1995 (Thomassen and Dawes 1996). The Qaanaaq region was the focus of further work by GEUS and the BMP (again funded by Greenland's government) in 2001. As in Upernavik, fieldwork was carried out to investigate the area's mineral occurrences and assess their potential for the mining industry. Despite the extensive cartographic mappings by geologists such as Koch, little attention had been given to understanding the mineralisation of this part of the old Thule district before the 2001 survey, although previous work that had been done included mineral exploration by the Greenarctic Consortium between 1969 and 1973, the GGU in 1975 and 1977, RTZ in 1991, and Nunaoil in 1994 and 1995 (Thomassen et al. 2002).

The Upernavik district was once more under the gaze of international oil companies from 2010, and Upernavik town was used as a base for offshore activities, prospecting, scoping projects and site surveys, and environmental assessment work. Extensive seismic surveys then took place in 2012 and 2013. The survey ships used the town's harbour facilities and a regular turnover of industry personnel and crews passed through the airport. There was heavy demand for accommodation, a number of private lodgings were established, and some smaller, local companies were contracted for the purposes of assistance with transport, supplies, and cargo. The oil companies and seismic survey crews also needed logistics support and this provided some local seasonal employment. For a small north Greenland town that was experiencing the negative effects of a changing climate on the marine environment as well as a decline in fish stocks at the time, there was considerable excitement – mainly on the part of municipal authorities and some local entrepreneurs – that Upernavik would become a permanent base for the oil industry on Greenland's emerging northern resource frontier. A plan for the town's development included a new harbour and related oil industry infrastructure and facilities as well as new housing.

The intense seismic campaigns did not come up with any data that indicated the presence of oil, and no further exploratory activity has been carried out since the surveys of 2013. Initially, this had a noticeable impact on a few local businesses, especially those that had pinned considerable hope on an oil boom, but also on how people feel about future oil development. While Greenland's government remains committed to developing a mining industry, the decision made to suspend oil exploration in 2021 has caused some to doubt that the seismic vessels or exploration ships will ever return to northern Baffin Bay. To the north of the harbour and the edge of the town's housing area, a large storage and construction site, with shipping containers, vehicles, scrap, other oil company equipment, a half-finished helicopter pad, and a larger concrete wall symbolise both the failure and hope of this unfinished project of hydrocarbon exploration and municipal development. Avannaata Kommunia's town plan for Upernavik, however, still has further development of the harbour, industry, infrastructure, and housing in mind, in the event of greater marine traffic and mineral exploration in the High

Arctic and the possibility of the Northwest Passage being used more as a regular, established maritime route by cargo vessels and cruise ships.

There were some in the town who had also thought that oil would provide full-time employment. In summer 2014, I spoke with a friend I have known since the late 1980s – I will call him Jens here – about what he thought about oil exploration and the seismic surveys. Originally from Kangersuatsiaq, Jens had moved to Upernavik three years previously because the fish processing facility in the settlement had been closed by Royal Greenland because there was a lack of fresh water needed for operating it and for processing and freezing Greenland halibut. Since being in Upernavik, however, Jens had struggled to find work. He had not moved there because the conditions for hunting and fishing were any better. Indeed, it was widely known that many hunters in the town had already given up their dog teams because the sea ice around the island was no longer forming as people expected it should usually do. With possibilities for hunting and fishing diminishing and being challenged because of these changing environmental conditions, they could no longer afford to keep and feed their dogs, so as many as possible were given to hunters in the district's settlements or had to be put down. But hunters throughout the district – not just in the town – were encountering and experiencing far more difficult ice conditions, and they remarked on the extreme weather events, the warmer winters, the shifting currents, more powerful waves, stronger winds, wetter summers, and fiercer storms that were becoming a part of daily life. This was adding to the time and effort involved in hunting animals as well as bringing greater danger and risk when travelling on both ice and open water.

Instead, having given up a life as a hunter and fisher in Kangersuatsiaq, Jens found part-time work for a while with the municipality driving the water delivery truck, but he heard there would be opportunities in constructing the new harbour for the seismic survey vessels and the oil rig supply ships that would crowd Upernavik's bay as some imagined it in the future. He told me he thought that the oil industry, while not necessarily desirable, would be a good thing for the town as it would provide local people with jobs. The year Jens moved to town, though, was also when, in retrospect, the vision of the Baffin Bay oil boom was beginning to fade. In the years that followed, when I saw Jens again on my visits north, he was still picking up part-time and casual work with the municipality, but he had also tried to make a living by selling his own carvings to tourists who visited Upernavik on cruise ships during the short summer. His work was beautiful and stunning, depicting animals shape-shifting into humans and vice versa. He made smaller pieces of jewellery too – earrings and necklaces. A problem Jens encountered, however, was that he used a lot of ivory from narwhal tusks and its export is prohibited by CITES, which means tourists cannot buy it and take it out of Greenland.

While Jens may have thought the arrival of the oil industry was going to usher in a prosperous time, I have spoken with many other people in the Upernavik and Melville Bay area – in their homes, in their winter and spring hunting camps on the sea ice, and in their summer fishing camps – who thought otherwise, and they remain worried that the oil companies will indeed be back some day, especially if a future government decision in Nuuk allows drilling again. However, current concerns over extractive industry mainly arise from mining projects near Savissivik and south of Upernavik town in parts of Sigguup Nunaa as well as in the Avanersuaq area and the increased activity and shipping this will entail. In 2020, British multinational Anglo American was granted a five-year exploration licence in Sigguup Nunaa and in a large area to the east of Kangersuatsiaq and Upernavik Kujalleq, for nickel, copper, and platinum group metals (PGM's), which include ruthenium, rhodium, palladium, osmium, iridium, and platinum. Even if oil exploration may not happen again for some time, I mentioned in Chapter 3 that, in September 2021, GEUS carried out a survey of some 800 kilometres of seabed off Greenland's west coast near Maniitsoq as a first phase in determining the presence of marine diamond deposits. The survey had been commissioned by De Beers and suggests that Greenland's government may not consider the marine environment to be off limits entirely to other kinds of investigation by extractive industry – in this case, for the prospects of seabed mining. When people in the Upernavik district heard about the survey, they thought back to the seasons when the seismic survey vessels were carrying out their intensive campaigns.

Based on the experience of those seismic surveys, hunters and fishers talk about the necessity of communities being prepared to deal with industry on an increasingly regular basis. In previous writing, I discussed how a hunter from the community of Nutaarmiut, which is to the north of Upernavik town, told me in summer 2015 that he anticipated the return of the seismic vessels and the oil company consultants: "The most important thing is to have consultation," he said at the time. "The oil companies have to come and talk with the hunters in the settlements. It is really important to combine knowledge before *anything* starts. It is important to develop the respect and relationships right from the beginning" (Nuttall 2017: 189). I described how he had spoken, like many other people I know in the area have done as well, of the frustrations they felt when the oil companies had first arrived in the district to talk about their plans. Visits to communities were short, and people said that they had received little information about exploration and the nature of seismic surveys. Instead, what the company delegations did bring with them was a plentiful supply of fresh fruit to distribute during the information sessions. They felt that the company executives and consultants had no interest in understanding the nature of the environment and the importance of animals for local livelihoods. Now, several years after these conversations I had with people in Upernavik, nothing seems to have changed. In particular, considerable concern is still expressed by hunters about the effects of mineral and oil exploration on the marine environment, as well as what mining companies may be doing on land. They continue to talk about the lingering effects of the seismic surveys on narwhals, which are hunted mainly during the open-water season from August to September – the same period when seismic activities were taking place in the area.

Following the surveys in both 2012 and 2013, hunters from communities in the Upernavik district, as well as from Savissivik, reported that narwhal behaviour was different. Indeed, and as I have reported on before, some felt that the hunt had been influenced negatively by the seismic activities (Nuttall 2016, 2017, 2019b). Since then, local observations have continued to indicate that narwhals remain restless and agitated since the seismic survey vessels were operating out at sea, even if it was a few years ago now. Each year, people remarked on how narwhals were seen to be moving closer to the coast, swimming deeper into ice-choked fjords and inlets (which increases the risk of ice entrapment for narwhals when the sea eventually freezes in autumn or early winter).

In May 2017, I discussed some of these concerns with hunters in Kangersuat-siaq, a community to the south of Upernavik town – and which I mentioned in Chapter 1 was established by the KGH as Prøven in 1800. It is a place in which I first carried out fieldwork in the late 1980s. We talked at length, as we had done on several occasions since 2014, about how the seismic surveys *pikitsip-paa* (makes alarmed or agitated) the narwhals. There are also other words used by hunters in Kangersuatsiaq and in other communities along this part of the northwest coast to describe narwhal behaviour since the seismic surveys were conducted. *Katsungaarpoq* refers to how narwhals are restless or in a hurry (the opposite, *katsorpoq*, means to be calm); some narwhals have been described by hunters as *eqqissinngilaq*, which means not to have peace within one's self or not to be left at peace; hunters also observe that narwhals are sometimes confused or perplexed because they are frightened of something (*uisanguserpoq*). None of these words, they report, were used to describe narwhal behaviour before the seismic vessels were operating. Places on land and sea can themselves be agitated, according to local perspectives, just as marine mammals and fish are. Sea ice, for example, can break off and suddenly go adrift because it is agitated, surprised, and fearful (*siku uippoq*). Hunters say that following seismic activities, they also noticed ice breaking off to a greater degree in areas of ice cover that were close to the coast.

While environmental changes in the marine ecosystem, such as thinning and declining sea ice, changing water temperatures, and changes in the migration and distribution of fish also likely affect and influence narwhal movement (Greenland halibut are a key prey species for narwhals), hunters say they understand that narwhals are not necessarily agitated and disturbed by such changes alone. They are also aware of how difficult it is to identify a single cause for the changes in narwhal behaviour, where they go to feed, and their shifting migratory routes. The general consensus among hunters is that there are many observed changes in narwhal movement, sea ice, weather patterns, and so on, but that there is a multiplicity of factors at work that contribute to this (Nuttall 2017).

Although the seismic vessels have not yet reappeared in these northern waters, the region has seen an increase in activities related to the exploration and survey-ing of mining potential, such as the Anglo American investigations. During my fieldwork, people have often told me that they feel excluded from decision-making processes surrounding plans for extractive industry projects, and that they have no opportunity to discuss the ways and nature of *pinngortitaq* and animals with consultants. More often than not, as I have already mentioned, they do not re-ceive information about projects and the exploratory activities that take place. Much discussion about this in Northwest Greenland focuses on a desire to see greater emphasis on the inclusion of local knowledge and local observations of change in social and environmental impact assessments.

Indigenous knowledge and impact assessments

A traveller who arrives at the town of Upernavik on the Dash-8 flight that connects it with Ilulissat and Qaanaaq may notice a large boulder located just within the airport's perimeter fence, close to the terminal building's entrance and a small parking area for vehicles. It is a big enough rock, but it may also go unremarked, for the airport is located almost at Upernavik's highest point. It is a very rocky area – there are many other boulders around – and the panorama also grab's one's attention. There is nothing immediately distinctive about this particular detached rock fragment. Although the airport is only at an elevation of 126 metres or 414 feet, in clear weather one walks out of the small terminal building to a dramatic view of mountains, islands, bays and inlets, the Upernavik ice fjord, and to the west, an iceberg-studded Baffin Bay. Given the grandeur, someone visiting Upernavik for the first time can be excused for not seeing the boulder or not paying too much attention to it even if they do notice it. However, like many things in the Upernavik area, it has a story that people tell in the town and throughout the district.

Upernavik is situated on a small island and in the late 1990s work began on the airport by levelling its mountain top. The only other possible location for an airstrip is on Akia, an uninhabited island separated from Upernavik by a narrow channel. For decades, and I recall this even being so in the late 1980s, Akia has been identified in regional development plans as the preferred and most logical site for an airport, but it would require the construction of a road and bridges to cross stretches of water. It was eventually decided that the mountain top location was the most cost-effective option, even though it is often shrouded in fog, and flights from Ilulissat and Qaanaaq are regularly cancelled or have to turn back if the runway is obscured by cloud cover. It was still a considerable engineering challenge. Some two years of blasting and construction – and disruption to everyday life in the town – followed and the airport was completed in 2000.

In one version of the story about the boulder, local people relate how, after a particularly intense day of blasting, the security guard who was keeping an eye on the airport construction site at night was visited by an *innersuaq* (plural *innersuit*; *innersuaq* means 'great fire' or 'great light', so perhaps one meaning of *innersuit* is 'great fire' or 'great light people'). *Innersuit* are human-like beings – some are smaller than humans, but they can also be the same size or even taller. They dwell underground, in one of the many non-human worlds of *Allanat* (Others), yet they live in ways that are similar to Inuit on the surface. They hunt and they fish and their subterranean world is one of abundance. However, they often come and go between the human and non-human domains, the boundaries between which are never completely solid anyway. Some stories reveal how an *innersuaq* can live with Inuit and marry into human families, while retaining their ties to the world of the *Allanat*. This night-time visitor at the airport site was much shorter than the guard. Pointing to the boulder, he told him that it had been dislodged by the explosives that day and had landed on his brother, killing him instantly. The *innersuaq* warned the security guard that humans could not move it, as it now marked his brother's grave, otherwise misfortune would follow. The weather would turn bad, he said, and there would be difficult times ahead for hunters and fishers.

There is a common thread in stories told about *innersuit* in Greenland. Misfortune can often arise if humans offend, injure, or kill them. They can turn good weather into foul weather, and they can break up sea ice and melt it away. But they can also bring good fortune and provide assistance and food to those in need, especially at times of starvation (e.g. Rink 1875; Sonne 2017). Many of the dramatic shifts experienced in Upernavik and the wider coastal region in sea ice resulting from a warming climate have occurred over the last twenty years. Hunting and fishing have suffered, and many people have experienced difficult times, as I indicated when I related the experience Jens has had (and see Nuttall 2019a, 2019b). The boulder has not been disturbed or moved since the airport construction, but sometimes people wonder about the coincidence nonetheless – after all, they say, the grave would still be affected by vibrations and noise from aircraft – and they reflect on whether the death of the *innersuaq* did indeed precipitate calamity and misfortune.

I relate this story about *innersuit* because in the work I do with communities in Northwest Greenland, I hear people talk about their concerns that the underground, which is also an *Allanat* world, is disturbed by seismic surveys and mining exploration, and *innersuit* often figure in their accounts of extraordinary happenings that coincide with the investigative activities of oil and mining companies. In 2017, I stayed with friends in Kangersuatsiaq who told me that for the past two or three summers, they had seen *innersuit*, who they thought would ordinarily live within the spaces along the coast and below the shoreline, moving inland to avoid disturbance from something that was happening out at sea. They attributed this to underground coastal *innersuit* experiencing continual shocks and sounds made by the airguns used by the seismic survey vessels as well as the entrances to their subterranean homes being lashed violently by the waves they thought must follow. My friends told me that they would regularly see *innersuit* family groups walking across the land – they described them as wearing skin clothing, carrying packs with their belongings and hunting equipment, and looking troubled and distressed. My friends also described how, at their fishing camp, they would hear rustling sounds and hushed voices at night, and found that fish would disappear from their drying racks and that some equipment such as fish hooks, lines, and nets would have gone. They felt the *innersuit* must be hungry to pilfer food and other items – but what explanation could there be, they asked, other than the animals and fish *innersuit*, like Inuit, depend on being scarce. The world within and below the ground must be disturbed, they thought, to the extent that animals are frightened and agitated.

Also in Kangersuatsiaq, I organised a community workshop in June 2015 to talk about climate change – the discussion turned to the seismic surveys and one hunter spoke about holes in the ground and crumbling rocks:

> In 2012, we noticed that there was warm water coming up from unidentifiable holes in the land, near the old school building here in Kangersuatsiaq. It

was at the end of February – it was cold, around -20 or -30°C, yet a lot of holes or geysers appeared with steam billowing from them. We had never seen it before. There was lots of steam coming from the warm water mixing with the cold. In 2013, there were scientists who came to study these things. But the people living here didn't get any information back about it. We've been noticing a lot of animals acting strangely or which seem to be frightened. We saw a harp seal without fur swimming nervously around an iceberg. There's got to be a connection because it was the exact same time seismic activities were going on.

In 2012–2013, at the same time the holes were observed, we noticed that some rocks had fallen into the sea. We found a huge hole where the rocks fell from the northeast part of the land. It was a circular hole and struck us as very odd. In Kangerlussuaq, we noticed parts of the slope of a big mountain crumbling away. We couldn't really see the source but the impact was certainly visible. A long time ago my father told me that he was out in the fjord and there was a helicopter travelling back and forth to the mountains. He heard sounds like explosions, as if there was some blasting going on. He had no idea what was happening up there because we never get information.

I restate a crucial point. While thought of as being empty of the traces of a long period of human occupation and contemporary social and cultural presence,

Figure 6.2 Kangersuatsiaq, Upernavik district (Photograph Mark Nuttall).

the areas that are defined as resource spaces are often lively places of multispecies encounters that are composed and brought into being through action and relations between the human and more-than-human, and between the intermingling between the worlds of Inuit and *Allanat*. As this chapter has discussed, oil exploration, for instance, has taken place in offshore waters that are critical for Greenlandic fisheries and for marine mammals, while many mining projects are close to communities or to the areas people hunt and fish in or travel through. Now that Anglo American has a licence to explore for nickel, copper, and platinum east and south of Kangersuatsiaq, its activities are taking place in a large area of customary use by Inuit in the area, in surroundings that are part of a wider memoryscape (Nuttall 1992). The Isua iron ore project in Nuup Kangerlua is a notable example of how Indigenous knowledge and experience can be elided from industry maps of extractive resource spaces. I have written before about how the social and environmental impact assessments described the area that would be marked off as the exclusive licence zone as a place that was far enough from human habitation and that social impacts would be negligible, despite its importance for hunting, fishing, and gathering activities for people from Nuuk and the small community of Kapisillit (Nuttall 2012). These resource spaces are also places described by scientists as ecologically unique but under threat (Christiansen, Mecklenburg and Karamushko 2014; Laidre et al. 2015; Reeves et al. 2014) or by archaeologists as having significance for cultural heritage (Myrup 2018). Again, such importance is often downplayed in assessments or the potential impacts are considered to be minimal.

Livelihoods and community economies in Northwest Greenland are not just bound up with animals, but are entwined with cryospheric elements too, as well as with the places in which *Allanat* dwell and make their own worlds. Absent from assessments, local people say, is a consideration of local knowledge about how such non-human entities compose, configure, and animate the world along with humans. A meaningful interest in local knowledge would require consultants to take time to listen to people about how they consider themselves and their lifeworlds to be entangled with the non-human and how human-environment relations are also shaped by political, economic, and cultural forces and processes. What is it, for example, that the movement, away from their subterranean abode, across the land of *innersuit* who appear distressed can tell us? How can this be recognised as vital local knowledge that can stand as testimony in an impact assessment? As Eduardo Kohn writes, such knowledge draws attention to the importance of recognising "a form of thinking about the world that grows out of a specially situated intimate engagement" (Kohn ibid.: 1) and "requires us to recognize the fact that seeing, representing, and perhaps knowing, even thinking, are not exclusively human affairs" (ibid.: 73). Being attentive to community perspectives on human-environment relations as well as Indigenous knowledge and insights on different forms of worldmaking by humans and non-humans; people, animals, *allanat*, glaciers, sea ice, among a diversity of multispecies – would be a major step forward in enacting impact assessments and conservation strategies that acquire local legitimacy as being inclusive and participatory. Beginning to grasp an understanding of the more than human and astonishing nature of the surroundings in which resource extraction projects will be carved out would allow for dialogue on the place of both the human and the more-than-human in a world of becoming, one that is characterised by extraordinary happenings.

References

Akerman, James R. 2009. *The Imperial Map: Cartography and the mastery of empire.* Chicago: Chicago University Press.

AMAP. 2012. *Arctic Climate Issues 2011: Changes in Arctic snow, water, ice and permafrost.* Oslo: Arctic Monitoring and Assessment Programme.

Andersen, Astrid Oberborbeck and Janne Flora. 2019. "Puzzling pieces and situated urgencies of climate change and globalization in the High Arctic." In Astrid B. Stensrud and Thomas Hylland Eriksen (eds.) *Climate, Capitalism and Communities: An anthropology of environmental overheating.* London: Pluto Press, pp. 115–132.

Axelsson, Ragnar. 2010. *Last Days of the Arctic*, with text by Mark Nuttall. Reykjavik: Crymogea.

Ball, Sydney H. 1923. "The mineral resources of Greenland." *Meddelelser om Grønland* 63 (1): 1–60.

Behn, Caleb and Karen Bakker. 2019. "Rendering technical, rendering sacred: The politics of hydroelectric development on British Columbia's Saaghii Naachii/Peace River." *Global Environmental Politics* 19 (3): 98–119.

Bobbette, Adam and Amy Donovan (eds.). 2019. *Political Geology: Active stratigraphies and the making of life.* Cham: Palgrave MacMillan.

Bøggild, Ove Balthasar. 1929. "Mining in Greenland." In M. Vahl, G.C. Amdrup, L. Bobé and AD.S. Jensen (eds.) *Greenland: The colonization of Greenland and its history until 1929.* Copenhagen: C.A. Reitzel, pp. 387–399.

Braun, Bruce. 2000. "Producing vertical territory: Geology and governmentality in late Victorian Canada." *Ecumene* 7 (1): 7–46.

Briner, Jason P., Lena Håkansson and Ole Bennike. 2013. "The deglaciation and neoglaciation of Upernavik Isstrøm, Greenland." *Quaternary Research* 80: 459–467.

Christiansen, Jørgen S., Catherine W. Mecklenburg and Oleg V. Karamushko. 2014. "Arctic marine fishes and their fisheries in the light of global change." *Global Change Biology* 20: 352–359. doi: 10.1111/gcb.12395.

Cowton, T.R., A. J. Sole, P. W. Nienow, D. A. Slater, and P. Christoffersen. 2018. "Linear response of east Greenland's tidewater glaciers to ocean/atmosphere warming." *Proceedings of the National Academy of Sciences* 115 (31): 7907–7912. doi: 10.1073/pnas.1801769115.

Delaney, Alyne E., Rikke Becker Jakobsen and Kåre Hendriksen. 2012. *Greenland Halibut in Upernavik: A preliminary study of the importance of the stock for the fishing populace.* A study undertaken under the Greenland Climate Research Centre. Aalborg University. Innovative Fisheries Management.

Deleuze, Gilles and Félix Guattari. 1988. *A Thousand Plateaus: Capitalism and schizophrenia* London: Bloomsbury.

Deleuze, Gilles and Claire Parnet. 2007. *Dialogues II.* New York: Columbia University Press.

Dodds, Klaus and Mark Nuttall. 2019 "Geo-assembling narratives of sustainability in Greenland." In Ulrik Pram Gad and Jeppe Strandsbjerg (eds.) *The Politics of Sustainability in the Arctic: Reconfiguring identity, space and time.* London and New York: Routledge, pp. 224–241.

Edney, Matthew. 2009. "The irony of Imperial mapping." In James R. Akerman (ed.) *The Imperial Map: Cartography and the mastery of Empire.* Chicago: University of Chicago Press, pp. 11–45.

Flora, Janne, Kasper Lambert Johansen, Bjarne Grønnow, Astrid Oberborbeck Andersen and Anders Mosbech. 2018. "Present and past dynamics of Inughuit resource spaces." *Ambio* 47, supplement 2, pp. 244–264.

Fredenslund, Theresa. 2022. "Identifying overexploitation in the coastal Greenland halibut fishery in the Disko Bay using static bioeconomic modelling." *Fisheries Research* 254, 106417.

Grønnow, Bjarne and Mikkel Sørensen. 2006. "Palaeo-Eskimo migrations into Greenland: The Canadian connection." In Jette Arneborg and Bjarne Grønnow (eds.) *Dynamics of Northern Societies*. Proceedings of the SILA/NABO Conference on Arctic and North Atlantic Archaeology, Copenhagen, May 10th-14th 2004 (Publications from the National Museum. Studies in Archaeology and History 10), Copenhagen: National Museum of Denmark, pp. 59–74.

Hansen, Keld. 2008. *Nuussuarmiut: Hunting families on the big headland*. Meddelelser om Grønland 35, *Copenhagen: Commission for Scientific Research in Greenland*.

Hastrup, Kirsten, Astrid O. Andersen, Bjarne Grønnow, and Mads Peter Heide-Jørgensen. 2018. "Life around the North Water ecosystem: Natural and social drivers of change over a millennium." *Ambio* 47 (Suppl. 2): 213–225.

Hastrup, Kirsten. 2019. "A community on the brink of extinction? Ecological crises and ruined landscapes in Northwest Greenland." In Astrid B. Stensrud and Thomas Hylland Eriksen (eds.) *Climate, Capitalism and Communities: An anthropology of environmental overheating*. London: Pluto Press, pp. 41–56.

Haubner, Konstanze, Jason E. Box, Nicole J. Schlegel, Eric Y. Larour, Mathieu Morlighem, Anne M. Solgaard, Kristian K. Kjeldsen, Signe H. Larsen, Eric Rignot, Todd K. Dupont and Kurt H. Kjær. 2018. "Simulating ice thickness and velocity evolution of Upernavik Isstrøm 1849–2012 by forcing prescribed terminus positions in ISSM." *The Cryosphere* 12: 1511–1522, https://doi.org/10.5194/tc-12-1511-2018.

Hayward, Philip. 2012. "Aquapelagos and aquapelagic assemblages." *Shima: The International Journal of Research into Island Cultures* 6 (1): 1–10.

Hayward, Philip. 2015. "The Aquapelago and the estuarine city: Reflections on Manhattan." *Urban Island Studies* 1: 81–95.

Henriksen, Niels. 1985. "Systematic geological mapping in 1984 in central and western North Greenland." *Rapport Grønlands Geologiske Undersøgelse* 126: 5–10.

Herbert, Marie. 1973. *The Snow People*. New York: G.P. Putnam's Sons.

Klinger, Julie. 2015. "A historical geography of rare earth elements: From discovery to the atomic age." *The Extractive Industries and Society* 2 (3): 572–580.

Kohn, Eduardo. 2013. *How Forests Think: Toward an anthropology beyond the human*. Berkeley: University of California Press.

Laidre, Kristin L., Harry Stern, Kit M. Kovacs, Lloyd Lowry, Sue E. Moore, Eric V. Regehr, Steven H. Ferguson, Øystein Wiig, Peter Boveng, Robyn P. Angliss, Erik W. Born, Dennis Litovka, Lori Quakenbush, Christian Lydersen, Dag Vongraven and Fernando Ugarte. 2015. "Arctic marine mammal population status, sea ice habitat loss, and conservation recommendations for the 21st century." *Conservation Biology* 29 (3): 724–737.

Laidre, Kirsten, Twila Moon, Donna D.W. Hauser, Richard McGovern, Mads Peter Heide-Jørgensen, Rune Dietz and Ben Hudson. 2016. "Use of glacial fronts by narwhals (*Monodon monoceros*) in West Greenland." *Biology Letters* 12 (10): 2016045720160457

Malaurie, Jean. 1982. *The Last Kings of Thule*. London: Jonathan Cape.

Meire, Lorenz, John Mortensen, Patrick Meire, Thomas Juul-Pedersen, Mikael K. Sejr, Søren Rysgaard, Rasmus Nygaard, Phillipe Huybrechts and Filip J.R. Meysman. 2017. "Marine-terminating glaciers sustain high productivity in Greenland fjords." *Global Change Biology*, doi: 10.1111/gcb.13801.

Myrup, Mikkel. 2018. *Moriusaq Archaeological Survey Report 2018*. Nuuk: Nunatta Katersugaasivia Allagaateqarfialu/Greenland National Museum and Archives, 18pp.

Nuttall, Mark. 1992. *Arctic Homeland: Kinship, community and development in Northwest greenland*. Toronto: University of Toronto Press.

Nuttall, Mark. 2012. "The Isukasia iron ore mine controversy: Extractive industries and public consultation in Greenland." *Nordia Geographical Publications* 41 (5): 23–34.

Nuttall, Mark. 2016. "Narwhal hunters, seismic surveys and the Middle Ice: Monitoring environmental change in Greenland's Melville Bay." In Susan A. Crate and Mark Nuttall (eds.) *Anthropology and Climate Change: From actions to transformations*, London and New York: Routledge, pp. 354–372.

Nuttall, Mark. 2017. *Climate, Society and Subsurface Politics in Greenland: Under the Great Ice*. London and New York: Routledge.

Nuttall, Mark. 2019a. "Sea ice, climate and resources: The changing nature of hunting along northwest Greenland's coast." In Astrid B. Stensrud and Thomas Hylland Eriksen (eds.) *Climate, Capitalism and Communities: An anthropology of environmental overheating*, London: Pluto Press, pp. 57–75.

Nuttall, Mark. 2019b. "Icy, watery, liquescent: Sensing and feeling climate change on northwest Greenland's coast." *Journal of Northern Studies* 14 (2): 71–91.

Petersen, Robert. 2003. *Settlements, Kinship and Hunting Grounds in Traditional Greenland* Meddelelser om Grønland 27, Copenhagen: Danish Polar Centre.

Reeves, Randall R., Peter J. Ewins, Selina Agbayani, Mads Peter Heide-Jørgensen, Kit M. Kovacs, Christina Lydersen, Robert Sydam, Wendy Elliot, Gert Polet, Yvette van Dijk and Rosanne Blijleven. 2014. "Distribution of endemic cetaceans in relation to hydrocarbon development and commercial shipping in a warming Arctic." *Marine Policy* 44: 375–389. https://doi.org/10.1016/j.marpol.2013.10.005.

Ries, Christopher. 2002. "Lauge Koch and the mapping of North East Greenland: Tradition and modernity in Danish Arctic research." In Michael Bravo and Sverker Sörlin (eds.) *Narrating the Arctic: A cultural history of Nordic scientific practices*. Canton, MA: Science History Publications, pp. 199–234.

Rink, Hinrich. 1875. *Tales and Traditions of the Eskimo*. Edinburgh and London: William Blackwood and Sons.

Rose, Edward P.F., Judy Ehlen and Ursula L. Lawrence. 2019. "Military use of geologists and geology: A historical overview and introduction." *Geological Society, London, Special Publications* 473: 1–29. https://doi.org/10.1144/SP473.15.

Shen, Grace Yen. 2014. *Unearthing the Nation: Modern geology and nationalism in Republican China*. Chicago and London: The University of Chicago Press.

Sonne, Birgitte. 2017. *Worldviews of the Greenlanders: An Inuit Arctic perspective*. Fairbanks: University of Alaska Press.

Stafford, Robert A. 1984. "Geological surveys, mineral discoveries, and British expansion, 1835–71." *The Journal of Imperial and Commonwealth History* 12 (3): 5–32.

Stroeve, Julienne C., John R. Mioduszewski, Asa Rennermalm, Linette N. Boisvert, Marco Tedesco and David Robinson. 2017. "Investigating the local-scale influence of sea ice on Greenland surface melt." *The Cryosphere* 11: 2363–2381.

Thomassen, Bjørn and Peter R. Dawes. 1996. "Inglefield Land 1995: Geological and economic reconnaissance in North-West Greenland." *Bull. Grønlands geol. Unders* 172: 62–68.

Thomassen, Bjørn, Johanne Kyed, Agnete Steenfelt and Tapani Tukiainen. 1999. "Upernavik 98: Reconnaissance mineral exploration in North-West Greenland." *Geology of Greenland Survey Bulletin* 183: 39–45.

Thomassen, Bjørn, Peter R. Dawes, Agnete Steenfelt and Johan Ditlev Krebs. 2002. "Qaanaaq 2001: Mineral exploration reconnaissance in North-West Greenland." *Geology of Greenland Survey Bulletin* 191: 133–143. https://doi.org/10.34194/ggub.v191.5141.

Van As, Dirk. 2011. "Warming, glacier melt and surface energy budget from weather station observations in the Melville Bay region of northwest Greenland." *Journal of Glaciology* 57 (202): 202–220.

Van den Broeke, Michiel R., Jason Box, Xavier Fettweis, Edward Hanna, Brice Noël, Marco Tedesco, Dirk van As, Willem Jan van de Berg and Leo van Kampenhout. 2017. "Greenland ice sheet surface mass loss: Recent developments in observation and modelling." *Current Climate Change Reports* 3 (4): 345–356. doi.org/10.1007/s40641-017-0084-8.

Vaughan, Richard. 1991. *Northwest Greenland: A history.* Orono: The University of Maine Press.

Weidick, Anker. 1958. "Frontal variations at Upernaviks Isstrøm in the last 100 years." *Meddelser fra Dansk Geologisk Forening* 14: 52–60.

7 Conservation and Indigenous rights

This chapter builds on my discussion of the environmental, social, and economic effects of resource exploration in northern Greenland in Chapters 5 and 6. I reflect on how the sea, sea ice, glaciers, coastlines, and wildlife of the region have become sites and objects for new forms of environmental governance. I argue that these are shaped by ideas of unique and fragile ecosystems that are identified, marked out, and categorised as being under threat by climate change but also by extractive industry, increased shipping, and geo-strategic competition. An area of interest in some of my current research is how northern Greenland as well as Canada's Arctic Archipelago and the Arctic Ocean have become critical spaces for action and interventions by scientists, conservationists, and environmental organisations. All share a deep concern over what is happening as the Arctic transforms. Their efforts are well-intentioned in how they advocate for the protection of the last fragments of disappearing ice, for the establishment of wildlife preserves, and the introduction of stricter conservation measures for Arctic animal species that form the basis of Indigenous hunting and fishing livelihoods. However, some of these interventions and strategies for conservation apply particular logics that tend to be informed and legitimised by scientific approaches, observation, and monitoring, and valuations of wildlife and ecosystems that in the process often work to exclude Inuit communities and Indigenous knowledge.

This makes for what can often be a turbulent politicised atmosphere in which contested ideas of environmental governance abound, illustrating a diversity of approaches to conservation and ecosystem protection in the contemporary Arctic. Marine biologists, for example, argue that polar bears, narwhals, and other marine mammals have become vulnerable and endangered (e.g. Laidre et al. 2015). This may be so in some cases, but in the arguments they put forward for greater regulatory action by governments and international conservation and management bodies to safeguard marine mammals and fish under threat from climate change, resource development, and over-exploitation, they accuse local communities of unsustainable hunting practices which they say only have a commercial economic incentive (Heide-Jørgensen et al. 2020), and recommend the implementation of precautionary approaches to commercial fishing as ice melts and Arctic waters warm (Christiansen, Mecklenburg and Karamushko 2014).

DOI: 10.4324/9781003175421-8

Rapid glacial melt and retreat, disappearing sea ice, geopolitical interests, and the making of extractive frontiers inform oceanic and terrestrial imaginaries that gesture towards a way of thinking about open polar seas and Arctic lands that may be devoid of ice in the future. Conservation and environmental organisations now commonly frame the Arctic as a critical zone of planetary crisis – a region that is not only experiencing dramatic climate change, but has come under greater international scrutiny. Groups such as the World Wide Fund for Nature (WWF) and Oceans North have launched campaigns – which are underpinned and sustained by narratives of vulnerability, risk, and ruination – to protect threatened northern waters, last areas of ice, polynyas, polar bears, and narwhals. As I have discussed so far in this book, people living in northern Greenland's coastal communities are also caught up in an economic and geopolitical scripting about frontiers of potential and abundance, global interest in energy resources and critical minerals, of business opportunities in a 'New Arctic,' and about threats to global security.

The manner in which conservationists, environmentalists, and scientists frame their approaches to protecting the last fragments of disappearing ice and Arctic species under threat is being unsettled and troubled, however, by the ways Inuit are challenging such action in how they respond to the claims made about who is entitled to speak for the non-human. Inuit calls for action in dealing with climate change and for regulating shipping, commercial fishing, and industry are expressed in the form of initiatives for co-management, community-based monitoring, and Indigenous Protected Areas (IPAs) that recognise Indigenous knowledge. Discussion about these initiatives, it must be pointed out, does often take the form of collaborative dialogue with conservation groups – and to be fair, WWF and Oceans North do engage with and are often supportive of Inuit communities (Oceans North, a Canadian organisation, has opened an office in Greenland, staffed by Greenlanders, and collaborates with the Inuit Circumpolar Council [ICC]) – even if these are not always friction-free partnerships to forge and sustain. Rather, as I see it, a challenge is the acceptance and integration of diverse and often divergent perspectives. Conservation organisations often proceed from a perspective that privileges the environment and animals, identifying specific ecosystems as being at risk and elevating some species to the status of emblems of a changing Arctic. ICC and other Indigenous organisations, however, seek to ensure environmental governance and conservationist narratives take serious note of Indigenous needs, including rights to hunt and fish, and perspectives on water and ice, as well as Indigenous knowledge of human and non-human entanglements that animate, invigorate and bring Arctic environments into being.

Fragments

Goodman et al. (2016) ask how it is that we *know* about the environment. In suggesting one answer, they contend that we now increasingly know about it – or think we do – through different forms, processes, and aspects of the spectacle. In particular, a diverse mediascape produces images and ideas of spectacular environments, and following on from this, they identify a trend of offering solutions to ecological problems through spectacular environmentalisms. By this, they mean media spaces that are environmentally focused and permeate and transcend everyday life, but which become normative, moralised, and politicised. In Chapters 1 and 5, I drew attention to how Greenland has been written about and represented as a remote polar region, often thought of as being located at the edge of the world. Such representations have contemporary relevance in how they inform ideas of the spectacular.

In *The Great Ice Age and Its Relation to the Antiquity of Man*, a treatise first published in 1874 on ice sheets and glacial motion, geology, climate history, and the emergence and movement of human societies in Western Europe following the last ice age, James Geikie described Greenland as an icy, forbidding place. "Fast as the snows deepen and harden into ice upon the bleak wild of Greenland," he wrote:

> that ice creeps away to the coast, and thus from the frozen reservoirs of the interior, innumerable glaciers pour themselves down every fiord and opening to the sea. Only a narrow strip of coast-line is left uncovered by the permanent snow-field or mer de glace—all else is snow and ice.
>
> (Geikie 1874: 56)

In this frozen world, glaciers, icebergs, and sea ice were written about by Geikie and many others at the time as presenting a formidable barrier to movement and navigation for those non-Indigenous visitors to Greenland who ventured north mainly in spring and summer – the explorers, whalers, adventurers, and scientists who thought of the Arctic as an unknown place that needed to be mapped, its resources assessed, classified, and to be exploited. Ice appeared to be ever-present, immutable, and forbidding. This was how the Arctic was supposed to look. And it was supposed to inspire feelings of awe.

In the nineteenth century, Arctic waters such as Baffin Bay and Melville Bay and other marine areas of northern Greenland and Canada's Arctic Archipelago were also considered places of terror by the Scottish and English whalers, who entered its ice-filled stretches on their way to the Arctic whaling grounds of the North Water and Lancaster Sound, as well as by the European and North American explorers who sought to chart the Northwest Passage or ventured to Smith Sound and Nares Strait with a view to finding a route to the North Pole. Filled with ice floes and icebergs and with coastal tidewater glaciers dripping down to the sea from the edges of Greenland's inland ice (each a powerful non-human actant), Melville Bay was experienced as a site of risk, trauma, and disaster. Ice conspired with wind and currents to hinder the progress of ships, confuse the crew members aboard them, and thwart, confound, and disrupt the economic venture of whaling and the advance of geographic exploration. Discovery ships were usually more strongly reinforced than the whaling vessels and so were not at the same risk of being lost in the Arctic (the ice proved to be more of an inconvenience, slowing them down rather than forcing them to turn back), but whaling vessels became stuck in the sea ice with alarming frequency, often for several weeks, and many vessels were wrecked and sank (Nuttall 2017, 2022a). Writing about Melville Bay in *The Threshold of the Unknown Region* in 1873, Clements Markham described how whaling ships from Scotland and England were crushed "like walnuts," ice "tore into their sides," and there were scenes of "indescribable horror" as great ships were "converted into 'shattered fragments.'" In this way, the Arctic was cast as a place of peril; it was thought of as an unforgiving environment in which the skills, strength, and courage of explorers, adventurers, and whalers were tested. Most returned to tell their tales of struggling with the ice, but many died and the Arctic was also imagined as a place that swallowed and consumed – a vast, unknown space where expeditions could get lost and where terrible things happened.

Terror and horror rather than the spectacular may have been recurring themes in tales of northern voyages and polar expeditions, but narratives about the Arctic sublime also gripped the public imagination. They were one way, at that time, through which European and North American publics mainly came to know about the Arctic. The nineteenth-century mediascape, which was comprised of books about exploration, art, poetry, maps with blank spaces beyond charted coastlines and depicting an open polar sea, newspaper reports, and novels that were inspired by Arctic discovery, still contributed to the sense that this was a region filled with spectacles.

Other, mainly non-Indigenous, travellers in northern Greenland in the late nineteenth and early twentieth centuries often fashioned themselves as explorers and adventurers not just in remote, unknown, blank spaces, but in lost worlds. Knud Rasmussen (1921), for example, considered the glaciers along the coast of Melville Bay to be vestiges of the last ice age, while American Arctic scientist William Carlson, who studied winter air currents in Greenland in the early 1930s, journeyed by dog sledge to the Cornell and Giesecke glaciers in the northern part of the Upernavik district, and described them as the ramparts of an ice kingdom. Reflecting upon the experience, Carlson wrote of being witness to an "unforgettable drama of vast ice bulks losing themselves in softening shadows, a dreamy melting of night with day shadows, the scenes drifting back past us, changing, darkening; nostalgic drama of perfect theater" (Carlson 1940: 279). Sledging to the front of Cornell Glacier, Carlson felt the Arctic revealed itself as "desolate" and "dreadful in its immense sterility" (ibid.: 275). These descriptions are of bleakness, the wild, rugged, and disorderly, the sublime, the terrible, and the awe-inspiring, in what Rasmussen, and Carlson as well as many others who travelled in northern Greenland, thought of as a remote, icy world of unmapped emptiness.

At the same time as they were entering and moving through remoteness and along the northern polar edges of the globe, Rasmussen and Carlson had a sense that they were experiencing a disappearing world. Rasmussen had Danish and Greenlandic ancestry and had grown up in Ilulissat, and he spoke fluent Greenlandic, but nonetheless often wrote about the Arctic as remote and forbidding, but still majestic and casting its spell on him rather than from the perspective of someone dwelling in an Inuit homeland and relating to in terms of always being aware of it as *pinngortitaq*. He thought the glaciers of Melville Bay to be "extravagant," throwing "gleaming ice mountains out into the ocean," but reaching Humboldt Glacier (at 79° North, and the widest tidewater glacier in the Northern Hemisphere) proved a disappointment; he called it a "half-dead ice-stream – scarcely capable of reproduction" (Rasmussen ibid.: 59). Carlson wrote in a concerned way about Giesecke Glacier receding and affected by powerful tides that swept and undermined its front, so much so that it left a great basin of water at the head of the fjord that was empty of icebergs and other pieces of floating glacial ice (ibid.: 281). What was left were fragments of those "vast ice bulks" that he had seen elsewhere in the district.

Contemporary scientific research on the effects of climate change on the polar regions and recent works on global ecological crisis have unsettled scholarly understanding and public imaginaries about the nature and seemingly enduring presence of the world's icy places (e.g. Wadhams 2016). Today, feelings of terror and horror still run deep in how the Arctic is written about and in the way we come to know it, but not necessarily because of a fear of being trapped in the ice or getting lost in a great, unknown frozen region. Rather, we think we are getting to know it too well, as satellite images of melting ice and maps of new shipping routes across open northern waters dominate our diverse mediascape, as celebrities engage in environmental campaigns to protect the ice (cf. Goodman et al. ibid.), and as tourism offers an opportunity to witness climate change and Arctic melt first-hand at places such as Ilulissat Icefjord. Instead of gazing upon images of the Arctic that inspire terror as well as feelings of the sublime (although the two are often one and the same), we come to think of what is happening there as terrifying. We look on in horror at images of starving polar bears on thinning ice and at filmed sequences of collapsing and receding glaciers – they may even be viewed up close from a cruise ship or tourist boat. It is in the Arctic that we can be witness to the world approaching a tipping point and veering towards catastrophe (cf. Brown 2011; Conkling et al. 2011).

As Arctic ice melts, scientific scenarios and environmentalist anxieties inform narratives of decay and ruination. With the very iciness of the Arctic undergoing dramatic change, scientists continue to warn that in the not too distant future, glaciers, ice caps, ice sheets, and sea ice will be reduced to such an extent that they will exist merely as fragments left behind by the rupture climate change is bringing. In climate change scenarios and strategies for Arctic conservation, one of the more bleaker images is of a future in which these icy fragments are remnants of a vulnerable Arctic that has been ravaged and mutilated by human action and capitalism in the Anthropocene. As such, they represent the detritus and debris left behind by climate change and resource exploitation that will linger and float in Arctic seas in "the aftermath of the process of breaking apart" (Orban 1997: 6). In these scenarios, fragments of the last ice in the northern polar world are attributed a value, along with endangered Arctic wildlife such as polar bears and narwhals, as vital elements that fill, constitute, and bring into being new frontiers of conservation as well as nurturing and informing a spatial politics of disappearance.

In Chapter 6, I discussed some of the processes and effects of climate change that scientists are studying in northern Greenland. Over the last decade, a large number of research projects, often comprising international teams of scientists, have generated data that points to how the seasonal melt of parts of the Greenland inland ice has been increasing in summer (e.g. Ryan et al. 2018).[1] Studies have documented rapid and dramatic ice-mass loss of Greenland's inland ice that is resulting from oceanic and atmospheric forcing (e.g. Bevis et al. 2019; Mouginot et al. 2019), a greater summer melting process that is affecting its edges and surface areas (e.g. Box and Decker 2011; Box et al. 2022) and the alarming retreat of large outlet glaciers (e.g. Andresen et al. 2014; Carr, Vieli and Stokes 2013; Harig and Simons 2012; Trusel et al. 2018). While there is increasing ice discharge from some of its thinning and retreating marine-terminating glaciers (Enderlin and Howat 2013), large amounts of ice mass are turning to meltwater and are flowing away towards the coast from Greenland's interior ice as streams and rivers, contributing to global sea level rise (Haubner et al. 2018; Khan et al. 2013). Increased meltwater run-off from glacial fronts is affecting water temperature and circulation patterns as well as the formation, thickness, and break-up of sea ice cover (Briner et al. 2013), and meltwater has been found to contain traces of dissolved organic carbon, making the Greenland ice sheet a source of such organic carbon that scientists discovered a few years ago is entering the Greenland and Labrador seas (Lawson et al. 2014). Glaciers are melting and ice shelves are reducing in area in other parts of the circumpolar north too, especially in the Canadian High Arctic (e.g. Harig and Simons 2016; Pendleton et al. 2019). Scientists warn that a warming Arctic means that some circumpolar regions, as evident in northern Greenland, are becoming less icy, far more slushy, and more watery places, but some are also becoming drier and dustier, rainier, and at times, cooler.

Many scientists with long-term experience of researching and monitoring sea ice have also been warning for some time that the Arctic Ocean's perennial ice cover is approaching a tipping point and may disappear completely before 2050 (e.g. Stroeve et al. 2012; Wang and Overland 2009; Wilkinson and Stroeve 2018), while others argue that a tipping point has already been reached – and a threshold crossed – meaning that ice-free summers will likely be the norm from the middle of this century (e.g. Wadhams ibid.). Observations and monitoring carried out over the last decade suggest that parts of the Arctic Ocean and surrounding seas are beginning to resemble the North Atlantic in terms of climate, as warm water inflow affects sea ice variability and marine productivity (Årthun et al. 2012; Lind, Ingvaldsen and Furevik 2018; Polyakov et al. 2017). This denser, saltier water inhibits sea ice formation and further increases the heat content of high latitude oceans. Lind et al. (ibid.), for example, argue that rising temperatures and diminishing ice in the Barents Sea north of Norway and northwest Russia are pushing it towards a tipping point and a new climate regime. Drawing on climate data going back some fifty years, as well as hydrographic observations carried out between 1970 and 2016, they show that sea ice decline and loss has been transforming the Barents Sea from a transition zone between the Atlantic and Arctic oceans to an ecosystem that is more like an extension of the Atlantic. In a synthesis of sea ice data and other sources reaching back to 1850, Walsh et al. (2017) suggest there is no relatively recent historical precedent on a pan-Arctic scale for the minimum sea ice extent that has been observed and monitored so far this century.

Scientists have also expressed concern that the Arctic Ocean is acidifying (Di et al. 2017), while the presence of microplastics in sea ice (which also moves and releases them around northern waters) has been identified as a growing environmental hazard (e.g. Peeken et al. 2018). Microplastics are a reminder that things drift to and gather in the Arctic. Toxic substances and objects are highly mobile, and they accumulate and have a lively presence and materiality in the Arctic. Contaminants enter human and animal bodies through the food chain and linger in marine and terrestrial environments, while other pollutants and murky substances that have been buried for decades, seep and emerge from melting ice, permafrost, and the sea bed. Some are often legacies from Cold War infrastructure and transport, such as abandoned military bases like Camp Century or oil spills from shipping as well as originating not only from recent industrial activity at more southerly latitudes but from the Arctic itself (e.g. Colgan et al. 2016; Langobardi 2014; Nuttall 1998). For many Indigenous residents across the Arctic who depend on the living resources of sea and land, life at the ice edge, on eroding coasts, or on the tundra seems increasingly uncertain, precarious, and risky. Indigenous resilience is being tested. The region is vulnerable and at risk. The people who live in the hunting communities along the Greenland coast need to catch marine mammals and fish for their own consumption, for distribution, and for sale. They are seeking ways to adapt to these transitions and also to changes in the presence and abundance of the living resources they rely on for their survival.

In Chapter 3, I wrote about how it often seems that to be in Greenland is to be with dust. Joseph Amato writes that dust is amorphous and is "found within all things, solid, liquid or vaporous" and "comes to rest everywhere in nature and on the human body. The finest dust—dust that can enter the pores of human skin—comes to rest in the ocean's depths" (Amato 2000: 4). To this can be added the fact that dust is found on ice sheets and within glacial ice in Greenland, Antarctica, the Himalayas, the Andes, the Rocky Mountains, the Alps, and anywhere else in the world's glacial regions. Dust not only gets in one's eyes, nose, mouth, throat, and lungs, mineral dust accumulates and settles on the inland ice and emerges from it, embedded in the ice cores extracted by scientists, washed out at its edges and revealed by glacial retreat.

Understanding the history and nature of dust in the cryosphere is vital for the documentation of past climate change and for offering clues about how the climate system may be affected in the future. The dust in ice cores has been found to originate almost entirely from Asian deserts and arid areas, specifically Mongolia's Gobi Desert and China's Taklamakan Desert, and is transported for thousands of kilometres by westerly winds (Biscaye et al. 1997; Bory, Biscaye and Grousset 2003). Ice core dust records have temporal significance for understanding glacial processes and environmental change and scientists examine them to understand the timing of changes to the ice sheet margin and relative sea level over the last glacial cycle (Simonsen et al. 2019).

Cryoconite, which is sediment found on the surface of the ablation zone of the inland ice and its glaciers (the ablation zone is where summer surface ice melt exceeds the accumulation of snow in the winter), also contains windblown dust particles from Asian deserts, although scientists have written about how much of it has a more local or regional origin as well as from volcanic eruptions and industrial activity on the other side of the world (e.g. Biscaye et al. ibid.; Hammer et al. 1978). This dark matter and the micro-organisms found in the water that accumulates in cryoconite holes lower ice albedo and contribute to melt. Dust accelerates the process of unbecoming. The inland ice may have long been subject to processes of deposition and accumulation from wind-blown dust and the black carbon particles originating from anthropogenic activity occurring far away from the Arctic, but Greenland's melting glaciers are also a source of high latitude dust emissions, as sediment is exposed, flushed out, and enters the aeolian system, with impacts on terrestrial, cryospheric, and aquatic environments (Bullard and Mockford 2018; Tobo et al. 2019; Wientjes et al. 2011).

It was Alfred Erik Nordenskiöld who named these vertical cylindrical holes cryoconite – its original phrasing was a combination of Greek and Swedish words to give the meaning of 'ice dust'. They were discovered by his expedition to the inland ice in 1870 and he described them as holes in the ice that were filled with water, with a layer of grey powder, just a few millimetres thick, at the bottom. His team observed them under microscopes, noticing what looked like white transparent grains, black opaque grains, green crystals, yellow translucent particles, and a sandy trachytic material that they concluded did not originate in Greenland. Significantly, they identified these dark holes as absorbing more of the sun's heat than the white ice around them, and in the process, they realised that they were contributing to the process of melt on the ablation zone (Nordenskiöld 1872).

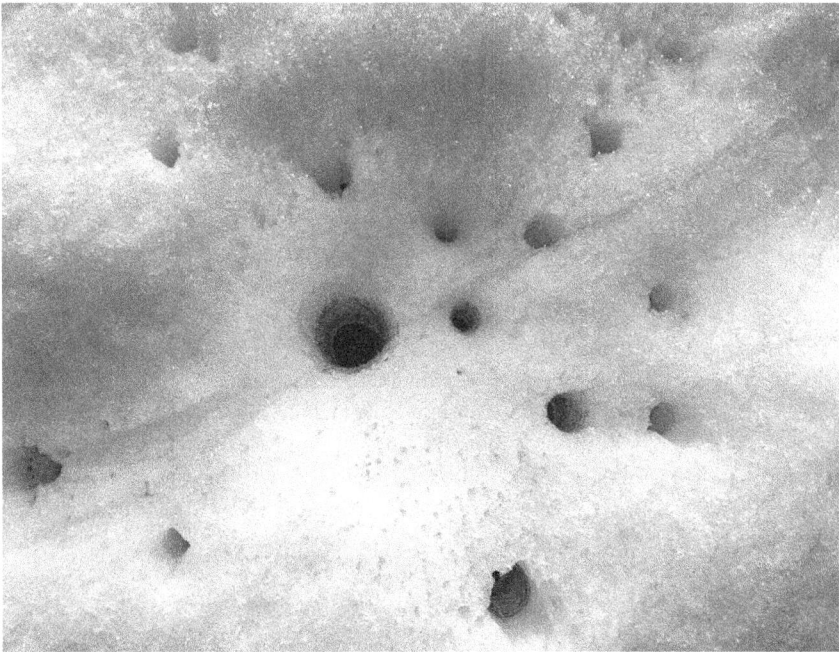

Figure 7.1 Cryoconite in the ablation zone (Photograph by Mark Nuttall).

To be in the ablation zone, to walk on the ice there, is to have an astonishingly direct encounter with cryoconite holes that seem to be everywhere (as well as with numerous ponds of meltwater). It also prompts deep reflection on how the granular and quite often, how the forms of matter that are unseen in the atmosphere, the ice, and the water have some of the most devastating impacts. Nieuwenhuis and Nassar (2018: 502) put it well when they observe how dust "picks up and leaves, shedding itself across the terrain it pervades," but how it can guide us "as we increasingly become more attuned to the indeterminacies and disorders of how we write an ever-fragmenting earth." Melting ice may be a powerful indicator of the effects of global climate change, but so is the presence of Asian dust on the inland ice. In Chapter 3, I also wrote about how dust seems to signify where things begin and where they end. As Amato (ibid.: 4) puts it, "Dust is everywhere because its source is everything." Beginning its windswept journey in Asia, dust settles on Greenlandic ice, gets embedded deep within it, and contributes to melt. Nordenskiöld traced how the organic material found in cryoconite eventually found its way to the edges of the ice. Identifying it as dried-up glacier streams, he described it in terms of fermentation, decay, and putrefaction.

To think about and to imagine what the Arctic will look and be like after (much of) the ice is gone – and when glaciers are inactive or dead – is, for many people, utterly disconcerting, extremely disquieting, and haunted by a sense of what will be absent and lost. But, as I have discussed in his book, the prospect of an ice-free Arctic provokes other kinds of interests, encouraged by ideas about economic opportunities and resource potential. The frozen northern seas of today are likely to be the open waters and new sea lanes of the not-too-distant future. While the nature, properties, and materialities of polar environments transform under the conditions and circumstances of climate change, many places across the Arctic are also being geo-assembled as sites of possibility and speculation for extractive industry and resource development (Dodds and Nuttall 2019).

Oil, gas, and mining companies and commercial shipping firms see climate change as allowing an expansion of maritime routes through northern waters, reducing the time a voyage takes via conventional passages such as the Suez and Panama Canals, enhancing global connectivity, enabling access to previously remote areas, and reducing sailing times between international ports and markets. Although the Northwest Passage in the Canadian Arctic has not yet emerged as a vital route for international shipping, a growing number of cruise ships do venture there (as they do in northern seas around Greenland, Svalbard, and Russia's Arctic islands), while coast guard vessels and the occasional commercial ship make summer voyages through these northern Canadian waters. Some cruise tour operators offer exciting itineraries for the intrepid tourist wishing to follow the routes of voyages of discovery of nineteenth-century explorers. A transit of the entire length of the Northwest Passage is no longer a rare event. Ironically, the idea of last chance tourism is also a draw for those keen to witness the changing Arctic – and experience ice, icebergs, polar bear sightings, and traditional Inuit culture before everything disappears.

Russia's Arctic waters present more likely possibilities for commercial shipping, however, especially given that the bathymetry of much of the Northwest Passage is not always well mapped in many places, especially in the central parts of Canada's Arctic region. Russia actively advertises and promotes its Northern Sea Route (NSR) as a shorter sea passage for transporting cargo between Asia and northern Europe than through the Suez Canal, and the country's more recent Arctic strategy has emphasised further development of this marine link as well as resource extraction in the north. Each year sees more ships travelling along Russia's Arctic coasts, from the Kara Sea to the Bering Strait, encouraged by rapidly changing sea ice and assisted by nuclear-powered icebreakers in enabling transit. Ironically, while climate change is contributing to the environmental conditions that allow more shipping to travel in Arctic waters, the shorter routes may lead to more efficiency in fuel consumption, thereby reducing maritime transport emissions. In October 2018, for example, the liquefied natural gas (LNG)-fuelled tanker *Lomonosov Prospect*, owned by the Russian shipping company Sovcomflot, completed its first commercial voyage along Russia's NSR to deliver a cargo of petroleum products from the Republic of Korea to northern Europe. The NSR section of the voyage from Cape Dezhnev at Chukotka to Cape Zhelaniya on the Novaya Zemlya archipelago took the vessel under eight days to complete, during which it covered a distance of 2,194 nautical miles. Such transit and the extraction of resources, while celebrated by industry, make scientists, environmentalists, and many residents of northern communities nervous about the further development of the circumpolar north, including the extraction of minerals and hydrocarbons. Large marine oil spills, for example, pose one of the major threats to the Arctic marine ecosystem. The risks associated with hydrocarbon exploration and production increase significantly with water depth, but also with other factors, such as sea ice, icebergs, storms, and winter darkness.

Melting ice may allow the smoother flow of shipping through northern waters and enable greater access to what have been regarded as previously remote Arctic regions to prospect, explore for, and extract minerals and hydrocarbons (as well as ship them to southern markets) but, as the previous chapter discussed, it impedes the mobility of Indigenous peoples such as those Inuit hunters and fishers who rely on sea ice and glacial ice for travel routes and on the ice-dependent animal species they catch. Retreating glaciers also have consequences for ecosystem productivity in Greenland's fjords, which, in turn, influences vital habitat for marine mammals and fish and affects hunting and fishing activities.

An extensive literature in conservation science has a concern with the fragmentation of ecosystems, the loss of wildlife habitat, and the diminishment of wilderness. Whether this focuses, for example, on the importance of forest fragments for biodiversity and species abundance for butterflies in the tropical urbanised landscape of Singapore (Koh and Sodhi 2004), or the vanishing grasslands of central California threatened by invasive plant species and suburban sprawl (Launer and Murphy 1994), to bats in the forests of Southeast Asia (Struebig et al. 2008), or primates in Africa (Marsh 2003), in environmentalist discourse the essential argument is that biodiversity is being ruined and destroyed by human action and the last surviving fragments need to be protected (Hecht, Morrison and Padoch 2014).

Fragments are residual. We can think of them as being artefacts or as the remains and remnants of larger objects that are left behind following the decline, disappearance, or end of a society, or following environmental ruin, or economic and industrial collapse. They can be intimations of past lives unearthed by archaeological excavation, for example, such as shards of pottery and glass, or tools, or pieces of tooth and bone, fossils or evidence of animal remains, fading strips of manuscript that shed light on a historical moment or event, traces of fences and walls, stone columns and foundations of buildings, or evidence of things that have been discarded, burnt, thrown away, and buried or submerged in middens and landfills (Harbison 2015). Fragments can also be left behind in the wake of a cataclysmic weather event or fire – dug from the rubble of people's homes, for instance – or as isolated stands of Douglas fir in temperate rainforests cleared by loggers' chainsaws. Fragments can also be extractions. And as Birch (2012) discusses with reference to the emergence of the bioeconomy, other kinds of fragments, such as tissue samples, genes, and stem cells, arise from new genetics technologies and are withdrawn with precision from living or dead bodies. Fragments are usually smaller and more scattered than other remnants. While remains can be visible in the form of ruins – of castles and medieval monasteries in Europe, for example – fragments are pieces that are sometimes much harder to unearth or bring to the surface and require some reassembling to get an idea of where they once fitted. As Sophie Thomas (2003: 181) puts it, fragments are "suspended between the part and the idea of the whole."

Ghostly species

As thinning sea ice, melting glaciers and ice sheets, and drifting icebergs enter the global imagination, they provoke anxieties over the disappearance of Arctic ecosystems and crucial wildlife habitat. Conservation and environmentalist action often hinge on ideas of loss and absence in a future Arctic that will be haunted by the spectral presence of iconic species such as polar bears that once roamed an extensive area of pack ice that will no longer exist in the coming decades. At the same time, polar bears, along with other charismatic polar wildlife such as whales and narwhals, become central to the most influential discourses about biodiversity and environmental change that shape regional and global approaches to marine conservation and resource development in the region. Jim Igoe (2017) argues that spectacles of nature act to variously connect or disconnect the experiences people have of their surroundings or places that are distant to them and he explores how images of threatened environments and endangered species are key to how people's perceptions of the world are shaped. They also have implications for the ways conservation can influence and disrupt human-environment relations. Similarly, Ursula Heise (2016) points to how the cultural production of endangered species has profound implications for the separation of the human and non-human in conservation discourse and practice.

Walker DePuy et al. (2022) call for reflection on and critical examination of the concepts and practices that are rendered visible and invisible in current environmental governance paradigms and pursuits and in what way they might advance or hinder the ability to envision and enact protected areas. They point out that many scholars, including Indigenous practitioners, have emphasised that the ideologies and enactments of coloniality are not only very much alive today in both knowledge and structural power relations, but that they live on through the dominant discourses, practices, and epistemologies of environmental governance. This echoes work by, for example, Mario Blaser (2016), whose research on how Innu in Labrador talk about caring for caribou points to ways of thinking about how a multiplicity of worlds that are animated in different forms troubles and challenges conservation ideologies and practices, and in the process, the relationships between Indigenous peoples and scientists – and how governance can fail if it ignores Indigenous ontology.

The narrative of exceptionalism that informs the geopolitics of and about the Arctic is reinforced in the ways ice, animals, and ecosystems are represented as threatened in global imaginaries about high latitudes (Dodds and Nuttall 2016). How the idea of spectral presence and absence can be extended beyond species to material substances such as ice is illustrated by Greenpeace's Save the Arctic campaign and by the WWF campaign to designate coastal stretches of northern Greenland and the Canadian Eastern Arctic and parts of the Arctic Ocean as the Last Ice Area. WWF argues that this area will be the only part of the Arctic marine environment to retain its summer sea ice by 2050. The Last Ice Area initiative aims to draw global attention to the loss of sea ice in a threatened ecoregion, but it is still a place, WWF argues, where sea ice and Arctic ecosystems are intact – at least for now, but they are threatened by climate change and industrial activity – and so urgent international action is needed to protect these last remaining stretches.[2]

It is in this way that the protection of ice has become key for ecological restoration in the Anthropocene Arctic. The Last Ice Area is described as a place that holds the promise for the formation and growth of new sea ice to replace that which has been lost. Calls for the protection of what remains of sea ice are not underpinned by ideas of novel ecosystems and the emergence of new species assemblages, however. While scientists warn that the rest of the Arctic Ocean is approaching tipping points and already crossing thresholds, WWF posits that protecting this last area of ice is the only chance left to reverse this shift. There is hope expressed that the loss of perennial sea ice is not irretrievable. The aim of such protection is to allow the Arctic Ocean an opportunity to resume its historic trajectory, and it is from the last remaining fragments that this may be possible.

The deployment of the image of the polar bear is key to the Last Ice Area campaign, especially in putting across a message that polar bears are threatened by extinction because of climate change. In doing so, the Last Ice Area campaign creates a particular wildlife-centric image of the Arctic in which people – most tellingly, Inuit hunters and their families who depend on catching and eating seals, whales, and polar bears – are relatively absent. Text, images, and films act together to construct, narrate, and disseminate a story that minimises and even ignores the presence of Indigenous people in the region while emphasising the iconic nature of charismatic wildlife to viewers outside of the Arctic. It is a form of spectacular environmentalism (Goodman et al. ibid.). The invocation of 'the' Last Ice Area has a certain power in how it appeals to governments and a global public by emphasising, with great urgency, that the region is in crisis and is undergoing rapid melt. It also implies that the area is the only one of its kind, and in doing so, it reinforces ideas of Arctic exceptionalism.

As historian Shane McCorristine (2018: 4) puts it:

> people from western cultures who visit the Arctic enter places that have long been imagined as somehow dreamlike or magical. 'Ice', 'wilderness' and 'sublime' register as keywords in a Eurocentric vocabulary that continues to inform the way that we think about what is Arctic and what is not Arctic.

McCorristone argues that these long-held associations contribute to the shaping of a public imagination in which the Arctic is enchanting, mysterious, magical, and even spectral. Its ruination demands global intervention to save it. In an insightful article McCorristine wrote with geographer Bill Adams (2020), attention is drawn to the implications of a sense of the spectral – of ghosts and haunting – for the way we think about conservation. Biodiversity conservation, they argue, is obsessed with the idea of unique species (notwithstanding debates about hybridity and interbreeding) that are either actually or almost extinct. They point to how the process of disappearing – and the state of disappearance – provides an immensely powerful frame for conservation thinking and action. They also discuss how the idea of extinction – the death of the last individual of a species – has repeatedly been used to describe the aspirations and work of the conservation movement. And so, they argue, extinction becomes a spectral event, in that a moment of loss, death, and disappearance generates lively, anxious, and provocative responses. These responses, they argue though, have the potential to be hopeful for the future and the survival of ecosystems and species as much as they are mournful about their loss and absence.

During my fieldwork, I have listened to local people in the Upernavik and Avanersuaq districts talk about how they feel WWF and the biologists who visit to study narwhals, polar bears, seals, and Greenland halibut are more focused on wildlife and the surrounding environment than the place of Inuit within these discussions. The hunters, with whom I work, tell me that they agree that protecting the ice on which polar bears, narwhals, and other animals as well as human communities depend is necessary. After all, and as I have discussed, the marine ecosystem supports their livelihoods and the unprecedented changes in the sea ice regime now being observed and experienced are having noticeable and problematic effects for hunting and fishing activities, for mobility, and for local economies. Yet they remain suspicious about the real intentions of organisations such as WWF as well as those of biologists from Nuuk, Denmark, and elsewhere, which many believe to be a concern with protecting animals with little regard for Inuit hunting and fishing rights. They have been critical at meetings held when WWF have visited communities to talk about the Last Ice Area initiative. Yet, while the Last Ice Area is gaining some political support, it is also entering mainstream scientific discourse as a space that is here and now, rather than a scenario of a possible future (e.g. Detlef et al. 2023; Moore et al. 2019).

Marine biologists, including those who do monitoring work at the Greenland Institute of Natural Resources and who provide the data the authorities in Nuuk need to decide on quotas, argue that narwhal hunting is not sustainable and that combined with the effects of climate change, local hunting practices threaten northern Greenland's sub-populations of narwhal (Heide-Jørgensen et al. ibid.). They call for a form of targeted conservation that would manage hunting or even place a moratorium on it, although hunters themselves argue that scientists direct their focus on harvesting activities without attempting to understand the cultural and economic importance of hunting, the nature of ice and the sea, the agency of narwhals, the ways they think and move around in the sea, and how people and narwhals in Northwest Greenland interact and interrelate. In many instances, those scientists who are tasked with acquiring data on narwhals reject the consideration that local perspectives and worldviews could be useful (Heide-Jørgensen et al. ibid.). With other anthropologists, I have been caught up in this debate and we have been accused of failing to understand the nature of the North Greenland ecosystem and serving as apologists for hunters, whose real incentive to go after narwhals, it is claimed, is commercial and market-driven, when we have suggested the importance of bringing science and local knowledge into conversation. It is a familiar clash between Indigenous knowledge, science, and the bureaucracies of wildlife management, which highlights a disjuncture between the epistemological and the ontological, between ways of knowing, thinking, acting, and being.

While WWF acknowledges the importance of ensuring sustainable communities in a changing Arctic, the worry for some people living in Northwest Greenland is that the Last Ice Area initiative could privilege ice and polar bears over Indigenous rights to hunt (Nuttall 2017). In their study of the Great Lebombo Conservancy in the borderlands of Mozambique, Massé and Lunstrum (2016) argue that the accumulation of securitisation is increasingly apparent in conservation practice. By this they refer to how security concerns related to environmental protection and wildlife management (and, in this case, poaching) take precedent over local people's rights and livelihoods. A result is that local knowledge is not only dismissed, but already vulnerable communities are dispossessed of rights and customary access to resources.

Domains of conservation and environmental management offered as a solution to Arctic meltdown and environmental crisis can be biopolitical, in the way that Biermann and Anderson (2017) consider approaches to the management of endangered species and protected areas to be an entanglement of logics and techniques that assert power, but are overlapping and contradictory and subject to multiple forms of biopolitics (Hodgetts 2017). They are also geopolitical in how ideas of the environment, endangered species, and biodiversity are deployed to influence the ways state governments and publics imagine and think about nature under threat. The non-human, natural resources, and the spaces in which they are found are defined and managed – often as eco-frontiers or enclaves, with governance structures that sustain their management – in ways that are not an outcome of local concerns, priorities, and decision-making, but of global processes and the work of transnational actors such as scientists and environmental NGOs who present scientific data and assert scientific facts (Ramutsindela et al. 2020).

Indigenous protected areas

The need to ensure the inclusion of Indigenous and local communities in conservation initiatives for water resources, ice, and wildlife acquires added urgency, given the speed of climate change and the major resource development projects being scoped out in northern Greenland as well as in neighbouring Arctic regions. In Nunavut, the Mary River project, a major iron ore mine currently in operation on Baffin Island, and the shipping of ore through Milne Inlet, east into Eclipse Sound along the southern edge of Bylot Island and south through Baffin Bay near Greenland's west coast to Rotterdam in the Netherlands is one significant example of this development. The first shipment of ore from the mine's port at Milne Inlet was in August 2015. In 2022, I heard from hunters in Mittimatalik (Pond Inlet) in northern Baffin Island as I travelled with them around Bylot Island and into the waters of Lancaster Sound at the mouth of Admiralty Inlet about how concerned they were about the mine's activities and the ships that carried the ore. However, they were also increasingly concerned about the greater number of cruise ships arriving in their community's waters and which they said made narwhals stay away. These concerns have also been expressed by hunters on social media.

Although the reduction of sea ice and open stretches of water during winter mean difficulties for hunters, and while increased meltwater runoff affects ecosystem productivity, in one part of the region it is actually the presence of more ice that poses one threat to the North Water Polynya, an area of open water that is essential to sustaining human and animal life. Polynyas are areas of usually open polar ocean surrounded by sea ice – if ice forms on the surface at all, it tends to be thin. The North Water is a unique High Arctic ecosystem that is largely ice-free throughout the year. However, it is vulnerable to climate change (Koerner et al. 2021; Ribeiro et al. 2021). Its ecological integrity is also threatened by resource development and increased shipping (for example, ilmenite from the mine at Moriusaq would be transported by ship through a stretch of the polynya). Scientists and conservationists are concerned about these changes and threats, but so too, as one would imagine they would be, are the people who depend on it.

The North Water Polynya is called Pikialasorsuaq in Greenlandic and Sarvarjuaq in the Inuktitut dialect of northern Nunavut. Both mean 'great upwelling.' The North Water was named as such by whalers in the nineteenth century, who were seeking ice-free passages and new hunting grounds for right whales and other species (Hastrup, Mosbech and Grønnow 2018). Its Indigenous name, however, captures its essence and very nature as perceived, observed, and experienced by those who live near it and who hunt and fish around its edges. It refers to how the mixing of water currents results in the upwelling of nutrients and so produces the attractive conditions and feeding opportunities favourable to marine mammals, fish, and birds (in Greenlandic, *pikialavoq* and *pikialaarpoq* mean 'to well out,' *pikippoq* means to be restless or to jump up, and *pikiarpoq* means a bird 'dives out of the water,' while *pikiarsaarpoq* means a seal 'dives out of the water').

Pikialasorsuaq/Sarvarjuaq is one of the most important marine ecosystems in Northwest Greenland and the eastern Canadian High Arctic and is the largest polynya in the Northern Hemisphere. Indeed, it is one of the most productive ecosystems of the planetary ocean. This ice-free area of rich biological diversity occurs annually and covers some 85,000 square kilometres. It depends on stable sea ice conditions around its boundaries and is characterised by a phytoplankton bloom in early spring. It sustains key Arctic species, including Arctic cod, seabirds, and marine mammals such as walrus, narwhal, beluga, polar bear, and several seal species, which are essential to the hunting and fishing economies of the Inuit communities in the region, in both Greenland and Nunavut (Flora et al. 2018; Nuttall 2019; Ribeiro et al. 2021). Hastrup, Mosbech, and Grønnow's (ibid.) description of it as a dynamic space is apt. They show how various physical, historical, and cultural processes and actions have intermingled, enhancing or obstructing each other and enabling or curtailing particular modes of human-animal relations.

The World Conservation Union has identified the North Water as one of the most ecologically significant marine areas in the Arctic and has proposed it as a candidate for a UNESCO Natural Marine World Heritage Site due to its outstanding global value and importance as an ecosystem (Speer et al. 2017). Yet,

Figure 7.2 At the ice edge near Qaanaaq (Photograph by Mark Nuttall).

changing ice conditions mean a vital ice arch – or an ice bridge – on its northern boundary in Nares Strait is less stable. Ice arches are patches of ice that form seasonally and prevent other pieces of ice from entering a body of water. They are critical to the formation of polynyas, but they break up – or collapse, as oceanographers describe it – in later spring and early summer. The collapse of the ice arch on the northern part of Pikialasorsuaq is becoming a trend in early spring, and this means the otherwise open waters of the polynya fill with pack ice sooner, with consequences for the phytoplankton bloom and the fauna dependent on it (Ribeiro et al. ibid.).

Research on the vulnerability of the North Water Polynya by conservation organisations such as WWF and Oceans North highlights a focus on marine ecosystems, while their support of work on the habitat of polar bears and ice as well as whales again foregrounds the prominence they put on charismatic megafauna when promoting the need for environmental protection. ICC has its own initiative under the Pikialasorsuaq Commission to which these organisations also contribute. Central to ICC's work is the argument that Inuit who live in the communities in the Pikialasorsuaq region are the ones who are best placed to set and lead a research agenda, to monitor change, and to establish their own realistic regulations for hunting and wildlife management that would allow for sustainable livelihoods and communities (Inuit Circumpolar Council 2017).

Collaborative decision-making processes and co-management arrangements between researchers, government agencies, and Inuit communities have been evolving since the 1970s (Berkes, Kislaliogu and Fast 2007). In Alaska and the Canadian North, Indigenous participation in resource management and decision-making on environmental matters is largely an outcome of land claims agreements. Community-based monitoring and co-management are effective tools for long-term Arctic monitoring systems. These systems help researchers understand the rapidly changing nature of northern environments, but they also empower northern communities and those who use and depend on animals and fish. In Nunavut, the Nunavut Land Claims Agreement (NLCA) sets out principles of conservation and Inuit harvesting rights, and Inuit communities and organisations work in collaboration with the Nunavut Wildlife Management Board (NWMB) and appropriate federal government departments towards co-management systems that integrate science and Inuit Qaujimajatuqangit (IQ). IQ is often translated to mean Inuit traditional knowledge, but more accurately it means "what Inuit have always known to be true" and captures both Inuit epistemology and ontology.

In November 2022, the US also set about strengthening these processes in collaboration with Indigenous peoples in Alaska and in all other states. The White House Office of Science and Technology Policy (OSTP) and the White House Council on Environmental Quality (CEQ) released a memorandum that recognises the importance of Indigenous Traditional Ecological Knowledge (ITEK) in federal scientific and policy processes and decision-making. The memorandum formally recognises ITEK as one of the many important bodies of knowledge that contributes to the scientific, technical, social, and economic advancements of the US and a collective understanding of the natural world. Here then is a commitment from the US government that Indigenous knowledge should inform the decision-making of federal agencies and departments.

Collaborative and community-based resource management is a decolonial approach that incorporates traditional knowledge and the observations and monitoring of Indigenous people as fundamental principles that should guide eco-system management. Following DePuy et al. (ibid.), done effectively, this would broaden the understanding of the need to think of ontological spaces in conser-vation, enhance collaboration, and centre the perspectives and work of those who have previously been excluded. Critical to this, as I pointed out in Chapter 6, is the need for an attentiveness to and recognition of Inuit understandings of the Arctic environment and human-animal relations. Scientific analysis of climate change in the Arctic – as much of the research cited earlier in this chapter illus-trates – tends to be rooted in categorisations of specific physical states and obser-vations of atmospheric processes and oceanic variability. This contributes to the dominant narrative about melting ice that has come to define the Anthropocene Arctic and how we worry about its future. True, the ice *is* melting, but in its concern with communicating the stark effect of the decline of sea ice cover and describing the rapidity of glacial ice discharge, surface runoff, and mass loss, this narrative frames one of the most apparent effects of climate change as liquifica-tion – or liquefaction – a phase transition from ice as solid to liquid. I would argue though that this does not capture the complexity of how Inuit living in North-west Greenland experience, in affective, sensorial, and immediately embodied ways, the shifting dynamics and volatile nature of their surroundings, how they relate to the other than human (including ice and water), how they experience weather and think about climate change, and how this informs ways of move-ment, anticipation, and adaptation on a daily basis.

I have pointed to Indigenous observations and scientific research that show how the reduction of sea ice cover and glacial mass are strikingly apparent. People in Northwest Greenland often talk in terms of how they feel their surroundings are being reconfigured and even disfigured by abrupt climate change and rapid melt. Once, camping at the ice edge for five days with three hunters from Qaa-naaq in March 2015, discussion centred on whether we should have gone fishing for Greenland halibut deeper in the fjord. We were not having any luck with spotting seals or walrus. They told me that March and April were really now the only months when the sea ice was best and when they could travel by dog sledge to the floe edge. But even as we were there, the ice was thin and we could hear it creaking and groaning as we lay in our tents. On the last day, one of the hunters shouted that we should move quickly – that we needed to pack the tents, load the sledges, and gather the dogs. The rest of us soon saw that the floe edge was break-ing up, and that our tents and sledges would soon begin to sink beneath its slushy surface. We did not linger. It has been on occasions like this that I have come to understand the ice edge as hunters now experience it – a place that has a far dif-ferent nature, texture and consistency than they say they have usually known it to be. It is a place of constant shifts, movement, and surprise. Its instability and fick-leness makes being there, camping and hunting, far more difficult and dangerous.

In the work I have done throughout Upernavik district, in communities from Kangersuatsiaq in the south to Kullorsuaq in the north, hunters and fishers I have interviewed have all spoken about how ocean currents are becoming more powerful and unpredictable and how unstable ice conditions are now commonplace and uncertain (see Nuttall 2017). However, while naturally concerned about climate change and its effects and while they have worries about economic conditions now and in the future, they do not necessarily talk about their surroundings in terms that convey a sense of environmental fragility or community vulnerability. Rather, a vernacular concerned with hunting, fishing, and travel in ever-changing surroundings expresses ways of thinking about fluidity, flexibility, and anticipation. It orientates people to living in a world of intentionality, action, agency, twists and turns, surprise, possibility, and choice; but it also prepares people for times that make them doubtful, unsure, uncertain, fearful, anxious, and apprehensive (see also Nuttall 2022b). Hunters and fishers in Northwest Greenland understand that when they set out to move and travel on sea, ice, and land, they must do so with awareness that the world around is capricious, fickle, emergent, and capable of throwing up surprises (Nuttall 2017, 2019); yet, the speed and extent of the changes they now observe and experience around them nonetheless presents considerable difficulties for travel, whether by dog sledge during winter and spring or by open boat during summer and autumn, and challenges their anticipatory knowledge.

Despite this, climate change is not necessarily experienced locally as liquification. Instead, I suggest that thinking of climate change as *liquescence* is a way to capture melt and thaw in those affective, sensory, and embodied ways in surroundings where the phased transitions of water as frozen, slushy, or liquid are not reflected upon in terms of scientific and ecosystem approaches to ice formation, melt, run-off, refreezing, and so on. Notwithstanding the uncertainties of being at the ice edge, for example, people – especially hunters and fishers – move through an environment anticipating encounters with icy, liquescent, and watery spaces often on the same day, whatever the season, and they sense how the weather is and how, through the movement of clouds or a change in wind direction, for example, it is likely to alter its mood. They anticipate such encounters just as much as when they are out walking around their home villages as they do when travelling by boat on the water or on the ice by dog sledge or snowmobile. They notice and feel these phased transitions on their faces, in their fingertips, in their toes, and in their bones, while they need to be sharply attuned to the navigational challenges of abrupt encounters with different conditions. As I have experienced on numerous occasions when travelling with hunters by dog sledge, for instance, headland cracks can suddenly open up on an otherwise smooth and apparently stable stretch of sea ice. Keeping dry is as vital as keeping warm – a constant preoccupation in surroundings that are increasingly moist, damp, rainy, and wet. To think of Northwest Greenland in terms of liquescence is to be reminded that changes in weather and climate are experienced as sensation as much as they are geophysical encounters.

Observing climate change from monitoring stations and sensing it remotely from space are critical to advancing our understanding of how the Arctic is being reshaped and reconfigured as ice melts, recedes, and shrinks (Comiso and Hall 2014), but this cannot possibly tell us anything at all about how flow, melt, moisture, and saturation are sensed, felt, and experienced by those who live and travel daily in surroundings that are increasingly liquescent. To describe the North Greenland environment in terms of water, ice, and land and assume they have distinct boundaries and properties only serves to reduce it to something far less intricate and patterned. Water, ice, and land overlap and intermingle – just as the places in which humans and non-humans, and Inuit and *Allanat* do. They merge and blur in different seasons and at different scales. They shimmer in different light when their edges and form are often difficult to determine. This is especially so in winter and spring, which are times when sea ice can play tricks on one's visual perception, confusing what appears to be terrain and surface, when snow cover can make land and ice appear indistinguishable, and when icebergs can appear to be islands or vice versa.

Conclusions

This chapter has illustrated how the waters, ice, and coastlines of Northwest Greenland have become sites and objects for new forms of conservation and environmental governance that are increasingly informed by ideas of them as unique, fragile ecosystems under threat. The nature, properties, and materialities of sea, ice, and land are changing, and quite rapidly so, but while concern is focused on melting ice, open seas, and warming waters, the High Arctic is also reimagined as a space of possibility and speculation, increasingly accessible to be made into resource spaces that can be prospected and from which minerals and hydrocarbons can be unearthed and amenable to encroachment and territorial claims. Conservation organisations frame the Arctic as a zone of climate change crisis and have launched campaigns – underpinned by narratives of ruination – to protect "last areas" of ice and ecologically sensitive waters such as the North Water Polynya. In this way, with its Last Ice Area campaign, WWF has defined parts of northern Greenland, High Arctic Canada, and stretches of the Arctic Ocean as an exceptional cryospheric space, one with globally vital ecosystems and wildlife habitat in need of protection. Marine biodiversity and iconic animals such as polar bears may be front and centre in their campaigns, but Indigenous rights and customary hunting and fishing practices are affected in the process.

However, Inuit responses are challenging and seeking to resist those environmental management and conservationist narratives that do not necessarily take note of Indigenous perspectives on place and the lively human and non-human relations that animate and bring Arctic surroundings into being. In doing so, claims are made for Indigenous knowledge to be brought into collaborative environmental governance, and this is illustrated through initiatives that call for community-based monitoring in places such as Pikialasorsuaq that would build the capacity for Inuit-led strategies for marine protected areas (Inuit Circumpolar Council ibid.). In Canada, the federal government supports initiatives for IPAs in the Arctic. Most notably, a new marine protected site called the Tallurutiup

Imanga/Lancaster Sound National Marine Conservation Area in the northern part of Nunavut was approved in 2017 as a National Marine Conservation Area. Federal and territorial agencies work with local Inuit communities and organisations to conserve the biodiversity of Lancaster Sound and its adjacent waterways. A similar initiative is also underway north and east of Ellesmere Island, Axel Heiberg Island, and other islands in Canada's Arctic Archipelago – called the Tuvaijuittuq Marine Protected Area, this stretches north into the Arctic Ocean, to the limits of Canada's maritime boundary. There is also Canadian government recognition of the vital importance of the North Water ecosystem (e.g. DFO 2021) and support for a similarly managed conservation area of Pikialasorsuaq/Sarvarjuaq that would be cross-boundary in nature and would connect Inuit communities in Nunavut and Greenland and allow for collaborative governance between the Canadian state and the Kingdom of Denmark.

Yet the Government of Greenland is not entirely convinced of the need to create such a special zone – or even convinced about its designation as a Marine World Heritage Site. And here we see the difference in scalar strategies between the Greenlandic authorities and the ICC that Gerhardt (2018) discusses. ICC's Pikialasorsuaq Commission is an example of how the ICC constructs an identity based on historical and contemporary common cultural bonds across the circumpolar north. Pikialasorsuaq is presented as a place that has sustained Inuit for millennia and which today connects Inuit communities across northern Baffin Bay. In meetings I have had with ICC representatives in Nuuk, they have emphasised that Pikialasorsuaq is a place that connects in temporal, spatial, and cultural ways – it is shared place that is vital for Inuit history and heritage, movement, mobility, and contemporary and future livelihoods. Greenlandic nation-building, however, is about "the construction of a bounded national identity, seeking a territorially based state sovereignty" (Gerhardt ibid.: 115). In this ideological mapping, the Greenlandic nation and the emergent Greenlandic state is also more than surface terrain, ice, and coastal waters. It stretches below ground into the subsoil, the ocean depths and along its continental shelf, and reaches into the skies above. With state formation, Greenland's territorial extent is still being determined and is in a process of becoming as the country continues to be emergent and to be made. Efforts to establish protected areas and conservation zones that are built on a foundation of Inuit knowledge, historic and contemporary occupancy, and rights are aspirations for the recognition of Indigenous sovereignty that is seemingly inconsistent with a Greenlandic national identity that does not privilege any particular ethnic group. It would conflict with a Greenlandic national interest, especially one that cultivates the idea of an independent Greenlandic state being possible and national strategies that include mining as a critical pillar for economic development. Pikialasorsuaq as a protected place under an agreed convention would situate Greenlandic national space – and the emergent Greenlandic state – within an international arrangement that would involve multiple agencies representing and enforcing different levels of national, territorial, municipal, and regional and local interest, including an acknowledgement of the rights and knowledge of a diversity of Indigenous Inuit communities. It would also hinder resource projects, exploration, and shipping in northern Greenland. There is a regional dimension to the opposition to protected sites too, as some politicians in Ilulissat think of the Ilulissat Icefjord designation as a UNESCO World Heritage Site as something that will limit and prevent the expansion and development of the town as an Arctic hub. As Naleraq-affiliated member of the Avannaata Kommunia board Anthon Frederiksen argued in July 2022, the UNESCO-listed site imposes restrictions on urban development and the construction of new roads, and the business development opportunities the new airport would enable.[3] North Greenland's marine regions not only act as connectors of sites of potential resource extraction to markets and consumers, they also allow for the establishment of a range of infrastructure and a multiplicity of industrial ventures and sites that link the various nodes and centres that are embedding Greenland in the spatial dynamics and economic drivers of the planetary mine.

This book has argued that in Greenland today, the politics, economies, and technologies of extraction are being formed, shaped, mobilised, and enacted in resource spaces that are imagined by company executives, consultants, economists, and engineers as ontologically distinct from society. These spaces are often described as remote, wild edges and depicted as empty of human history and contemporary presence – rather than being infused, as Inuit see them, with their essence and traces. At the same time, they are assembled, constituted, and made as sites of potential that promise great returns on investment. Drawing on David Harvey (2006), these sites can be considered spaces of global capital that are accessible to the techniques, technologies, and labour necessary for extraction to occur. In the erasure of the social and the cultural traces of long occupation or movement, along with the environmental rupture that follows extraction, resource spaces also take shape and are given value through a process of "accumulation by degradation" (Johnson 2010). Ironically, while rendered as empty of human life, resource spaces are abstracted by geophysical analysis of rock, technical mappings or layers of ore, political discourses, economic forecasts, and business assessments. They are re-animated as lively sites of extraction that will provide many of the material conditions for a post-colonial, cosmopolitan, and independent Greenland that is in the making (Nuttall 2017). These abstractive ideas and practices extend to the making of new frontiers of conservation as much as they do resource spaces.

The subterranean may have become far more vital to Greenlandic aspirations for state formation and greater autonomy in a rapidly changing part of the world, but global interest in what lies underground has also framed the country's geopolitical position and continues to do so with added urgency as the world looks to regions such as the Arctic for critical minerals to enable new technologies for climate change solutions. The obvious irony here is that pursuing sustainability through a green energy agenda and mitigating the effects of climate change nonetheless requires the intensification of mining those minerals. No matter what is being unearthed and extracted from the underground or the seabed, it is difficult when developing a mine to avoid ecological and social disruption, even if that disruption is assessed, as some projects claim, as negligible. Once unearthed, minerals are gone for good. And once unearthed, they have to be moved and transported by land and sea for processing and production. It is to be hoped though that Inuit perspectives on place and human and non-human entanglements can counter ways in which parts of Greenland are abstracted and represented variously as empty frontier zones and wilderness areas (yet potentially rich in subsurface resources), geopolitical spaces, or as threatened, disappearing regions of ecological and species uniqueness.

Notes

1 The National Ice Snow and Data Center at the University of Colorado, Boulder, provides daily satellite images and information about melting on the Greenland ice sheet: http://nsidc.org/greenland-today/.
2 See WWF's Last Ice Area website: https://www.arcticwwf.org/about/the-arctic/arctic-regions/last-ice-area/.
3 "Naleraq vil droppe UNESCO i Ilulissa" ("Naleraq will drop UNESCO in Ilulissat"), *Sermitsiaq* 14 July, 2022, https://sermitsiaq.ag/node/238441

References

Amato, Joseph. 2000. *Dust: A history of the small and the invisible*. Berkeley: University of California Press.
Andresen, Camilla S., Kristian K. Kjeldsen, Benjamin Harden, Niels Nørgaard-Pedersen and Kurt H. Kjær. 2014. "Outlet glacier dynamics and bathymetry at Upernavik Isstrøm and Upernavik Isfjord, North-West Greenland." *Geological Survey of Denmark and Greenland Bulletin* 31: 79–82.
Årthun, M., T. Eldevik, L.H. Smedsrud, Ø. Skagseth and R.B. Ingvaldsen. 2012. "Quantifying the influence of Atlantic heat on Barents Sea ice variability and retreat." *Journal of Climate* 25 (13): 4736–4743.
Berkes, Fikret, Mina Kislalioglu Berkes and Helen Fast. 2007. "Collaborative integrated management in Canada's North: The role of local and traditional knowledge and community-based monitoring." *Coastal Management* 35 (1): 143–162.
Bevis, Michael, Christopher Harig, Shfaqat A. Khan, Abel Brown, Frederik J. Simons, Michael Willis, Xavier Fettweis, Michiel R. van den Broeke, Finn Bo Madsen, Eric Kendrick, Dana J. Caccamise II, Tonie van Dam, Per Knudsen and Thomas Nylen. 2019. "Accelerating changes in ice mass within Greenland, and the ice sheet's sensitivity to atmospheric forcing." *PNAS* 116 (6): 1934–1939.
Birch, Kean. 2012. "Knowledge, place and power: Geographies of value in the bioeconomy." *New Genetics and Society* 31 (2): 183–201.
Biermann, Christine and Robert M. Anderson. 2017. "Conservation, biopolitics and the governance of life and death." *Geography Compass* 11 (10): e12329. https://doi.org/10.1111/gec3.12329.
Biscaye, P.E., F.E. Grousset, M. Revel, S. Van der Gaast, G.A. Zielinski, A. Vaars and G. Kukla. 1997. "Asian provenance of glacial dust (stage 2) in the Greenland Ice Sheet Project 2 Ice Core, Summit, Greenland." *Journal of Geophysical Research* 102 (C12): 26,765–26,781.
Blaser, Mario. 2016. "Is another cosmopolitics possible?" *Cultural Anthropology* 31 (4): 545–570.
Bory, Aloy J.-M., Pierre E. Biscaye and Francis E. Grousset. 2003. "Two distinct seasonal Asian source regions for mineral dust deposited in Greenland (NorthGRIP)." *Geophysical Research Letters* 30 (4), 1167. doi.org/10.1029/2002GL016446.

Box, Jason E. and David T. Decker. 2011. "Greenland marine-terminating glacier area changes: 2000–2010." *Annals of Glaciology* 52 (59): 91–98.

Box, Jason E., Alun Hubbard, David B. Bahr, William T. Colgan, Xavier Fettweis, Kenneth D. Mankoff, Adroen Wehrlé, Brice Noël, Michiel R. van den Broeke, Bert Wouters, Anders A. Bjørk and Robert S. Fausto. 2022. "Greenland ice sheet climate disequilibrium and committed sea-level rise." *Nature Climate Change* 12: 808, 813.

Briner, Jason P., Lena Håkansson and Ole Bennike. 2013. "The deglaciation and neoglaciation of Upernavik Isstrøm, Greenland." *Quaternary Research* 80: 459–467.

Brown, Lester R. 2011. *World on the Edge: How to prevent environmental and economic collapse.* New York and London: W.W. Norton & Company.

Bullard, Joanna E. and Tom Mockford. 2018. "Seasonal and decadal variability of dust observations in the Kangerlussuaq area, west Greenland." *Arctic, Antarctic, and Alpine Research* 50 (1). doi.org/10.1080/15230430.2017.1415854.

Carlson, William A. 1940. *Greenland Lies North.* New York: Macmillan.

Carr, J. Rachel, Andreas Vieli and Chris Stokes. 2013. "Influence of sea ice decline, atmospheric warming, and glacier width on marine-terminating outlet glacier behavior in northwest Greenland at seasonal to interannual timescales." *Journal of Geophysical Research: Earth Surface* 118: 1210–1226.

Christiansen, Jørgen S., Catherine W. Mecklenburg and Oleg V. Karamushko. 2014. "Arctic marine fishes and their fisheries in the light of global change." *Global Change Biology* 20: 352–359. doi: 10.1111/gcb.12395.

Colgan, William, Horst Machguth, Mike McFerrin, Jeff D. Colgan, Dirk van As and Joseph A. Macgregor. 2016. "The abandoned ice sheet base at Camp Century, Greenland, in a warming climate." *Geophysical Research Letters* 43: 8091–8096, doi:10.1002/2016GL069688.

Comiso, Josefino C. and Dorothy K. Hall. 2014. "Climate trends in the Arctic as observed from space." *WIREs Climate Change* 5 (3): 389–409.

Conkling, Philip, Richard Alley, Wallace Broecker and George Denton. 2011. *The Fate of Greenland: Lessons from abrupt climate change.* With photographs by Gary Comer. Cambridge, Mass. and London: The MIT Press.

DePuy, Walker, Jacob Weger, Katie Foster, Anya M. Bonanno, Suneel Kumar, Kristen Lear, Raul Basilio and Laura German. 2022. "Environmental governance: Broadening ontological spaces for a more livable world." *Environment and Planning E: Nature and Space* 5 (2): 947–975.

Detlef, Henrieka, Matt O'Regan, Christian Stranne, Mads Mørk Jensen, Marianne Glasius, Thomas M. Cronin, Martin Jakobsson and Christof Pearce. 2023. "Seasonal sea ice in the Arctic's last ice area during the early Holocene." *Communications Earth and Environment* 4, 86. https://doi.org/10.1038/s43247-023-00720-w.

DFO. 2021. *Proceedings of the Regional Peer Review on the Biophysical and Ecological Overview of the North Water Polynya and Adjacent Areas; January 22–24, 2020.* DFO Can. Sci. Advis. Sec. Proceed. Ser. 2021/011, Ottawa: Department of Fisheries and Oceans.

Di, Qi, Liqi Chen, Baoshan Chen, Zhongyong Gao, Wenli Zhong, Richard A. Feely, Leif G. Anderson, Heng Sun, Jianfang Chen, Min Chen, Liyang Zhan, Yuanhui Zhang and Wei-Jun Cai. 2017. "Increase in acidifying water in the western Arctic Ocean." *Nature Climate Change* 7: 195–199.

Dodds, Klaus and Mark Nuttall. 2016. *The Scramble for the Poles: The geopolitics of the Arctic and Antarctic.* Cambridge: Polity.

Dodds, Klaus and Mark Nuttall. 2019. "Geo-assembling narratives of sustainability in Greenland." In Ulrik Pram Gad and Jeppe Strandsbjerg (eds.) *The Politics of Sustainability in the Arctic: Reconfiguring identity, space and time,* London and New York: Routledge, pp. 224–241.

Enderlin, Ellyn M. and Ian M. Howat. 2013. "Submarine melt rate estimates for floating termini of Greenland outlet glaciers (2000–2010)." *Journal of Glaciology* 59 (213): 67–75.

Flora, Janne, Kasper Lambert Johansen, Bjarne Grønnow, Astrid Oberborbeck Andersen and Anders Mosbech. 2018. "Present and past dynamics of Inughuit resource spaces." *Ambio* 47, supplement 2, pp. 244–264.

Geikie, James. 1874. *The Great Ice Age and Its Relation to the Antiquity of Man.* London: W. Isbister.

Gerhardt, Hannes. 2018. "The divergent scalar strategies of the Greenlandic government and the Inuit Circumpolar Council." In Kristian Søby Kristensen and Jon Rahbek-Clemmensen (eds.) *Greenland and the International Relations of a Changing Arctic: Postcolonial paradiplomacy between high and low politics.* London and New York: Routledge, pp. 113–124.

Goodman, Michael K., Jo Littler, Dan Brockington and Maxwell Boykoff. 2016. "Spectacular environmentalisms: Media, knowledge and the framing of ecological politics." *Environmental Communication* 10 (6): 677–688.

Hammer, C.U., H.B. Clausen, W. Dansgaard and N. Gundestrup. 1978. "Dating of Greenland ice cores by flow models, isotopes, volcanic debris, and continental dust." *Journal of Glaciology* 20 (82): 3–26.

Harbison, Robert. 2015. *Ruins and Fragments: Tales of loss and rediscovery.* London: Reaktion Books.

Harig, Christopher and Frederik J. Simons. 2012. "Mapping Greenland's mass loss in space and time." *PNAS* 109 (49): 19934–19937.

Harig, Christopher and Frederik J. Simons. 2016. "Ice mass loss in Greenland, the Gulf of Alaska, and the Canadian Archipelago: Seasonal cycles and decadal trends." *Geophysical Research Letters* 43 (7): 3150–3159. doi: 10.1002/2016GL067759.

Harvey, David. 2006. *Spaces of Global Capitalism: Towards a theory of uneven geographical development.* London: Verso.

Hastrup, Kirsten, Anders Mosbech and Bjarne Grønnow. 2018. "Introducing the North Water: Histories of exploration, ice dynamics, living resources, and human settlement in the Thule Region." *Ambio* 47 (Suppl. 2): 162–174.

Haubner, Konstanze, Jason E. Box, Nicole J. Schlegel, Eric Y. Larour, Mathieu Morlighem, Anne M. Solgaard, Kristian K. Kjeldsen, Signe H. Larsen, Eric Rignot, Todd K. Dupont and Kurt H. Kjær. 2018. "Simulating ice thickness and velocity evolution of Upernavik Isstrøm 1849–2012 by forcing prescribed terminus positions in ISSM." *The Cryosphere* 12: 1511–1522. https://doi.org/10.5194/tc-12-1511-2018.

Hecht, Susannah B., Kathleen D. Morrison and Christine Padoch (eds.). 2014. *The Social Lives of Forests: Past, present and future of woodland resurgence.* Chicago and London: The University of Chicago Press.

Heide-Jørgensen, M.P., E. Garde, R.G. Hansen, O.M. Tervo, Mikkel-Holger S. Sinding, L. Witting, M. Marcoux, C. Watt, K.M. Kovacs and R.R. Reeves. 2020. "Narwhals require targeted conservation." *Letters, Science* 370 (6615): 416, doi: 10.1126/science.abe7105.

Heise, Ursula K. 2016. *Imagining Extinction: The cultural meanings of endangered species.* Chicago and London: The University of Chicago Press.

Hodgetts, Timothy. 2017. "Wildlife conservation, multiple biopolitics and animal subjectification: Three mammals' tales." *Geoforum* 79: 17–25.

Igoe, Jim. 2017. *The Nature of Spectacle: On images, money, and conserving capitalism.* Tucson: University of Arizona Press.

Inuit Circumpolar Council. 2017. *People of the Ice Bridge: The future of the Pikialasorsuaq. Report of the Pikialasorsuaq Commission.* Ottawa: Inuit Circumpolar Conference.

Johnson, Leigh. 2010. "The fearful symmetry of Arctic climate change: Accumulation by degradation." *Environment and Planning D* 28 (5): 828–47.

Khan, Shfaqat Abbas, Kurt H. Kjaer, Niels J. Korsgaard, John Wahr, Ian R. Joughin, Lars H. Timm, Jonathan L. Bamber, Michiel R. van den Broeke, Leigh A. Stearns, Gordon S. Hamilton, Bea M. Csatho, Karina Nielsen, Ruud Hurkmans and Greg Babonis. 2013. "Recurring dynamically induced thinning during 1985 to 2010 on Upernavik Isstrom, West Greenland." *Journal of Geophysical Research: Earth Surface* 118 (1): 111–121.

Koerner, Kelsey A., Audrey Limoges, Nicolas Van Nieuwenhove, Thomas Richerol, Guillaume Massé and Sofia Ribeiro. 2021. "Late Holocene sea-surface changes in the North Water polynya reveal freshening of northern Baffin Bay in the 21st century." *Global and Planetary Change* 206, 103642. https://doi.org/10.1016/j.gloplacha.2021.103642.

Koh, Lian Pin and Navjot S. Sodhi. 2004. "Importance of reserves, fragments, and parks for butterfly conservation in tropical urban landscape." *Ecological Applications* 14 (6): 1695–1708.

Laidre, Kristin L., Harry Stern, Kit M. Kovacs, Lloyd Lowry, Sue E. Moore, Eric V. Regehr, Steven H. Ferguson, Øystein Wiig, Peter Boveng, Robyn P. Angliss, Erik W. Born, Dennis Litovka, Lori Quakenbush, Christian Lydersen, Dag Vongraven and Fernando Ugarte. 2015. "Arctic marine mammal population status, sea ice habitat loss, and conservation recommendations for the 21st century." *Conservation Biology* 29 (3): 724–737.

Langobardi, Pam. 2014. "Plastic as shadow: The toxicity of objects in the Anthropocene." In Tricia Cusack (ed.) *Framing the Ocean, 1700 to the Present: Envisaging the sea as social space*. Farnham, Surrey and Burlington, VT: Ashgate Publishing, pp. 181–191.

Launer, Alan E. and Dennis D. Murphy. 1994. "Umbrella species and the conservation of habitat fragments: A case of a threatened butterfly and a vanishing grassland ecosystem." *Biological Conservation* 69 (2): 145–153.

Lawson, E.C., J.L. Wadham, M. Tranter, M. Stibal, G.P. Lis, C.E.H. Butler, J. Laybourn-Parry, P. Nienow, D. Chandler and P. Dewsbury. 2014. "Greenland Ice Sheet exports labile organic carbon to the Arctic oceans." *Biogeosciences* 11, 4015–4028, https://doi.org/10.5194/bg-11-4015-2014, 2014.

Lind, Sigrid, Randi B. Ingvaldsen and Tore Furevik. 2018. "Arctic warming hotspot in the northern Barents Sea linked to declining sea-ice import." *Nature Climate Change* 8: 634–639.

Markham, Clements R. 1873. *The Threshold of the Unknown Region*. London: Samson Low.

Marsh, Laura (ed.). 2003. *Primates in Fragments: Ecology and conservation*. New York: Springer.

Massé, Francis and Elizabeth Lunstrum. 2016. "Accumulation by securitization: Commercial poaching, neoliberal conservation, and the creation of new wildlife frontiers." *Geoforum* 69: 227–237.

McCorristine, Shane. 2018. *The Spectral Arctic: A history of dreams and ghosts in polar exploration*. London: UCL Press.

McCorristine, Shane and William M. Adams. 2020. "Ghost species: Spectral geographies of biodiversity conservation." *Cultural Geographies* 27 (1): 101–115.

Moore, G.W.K., A. Schweiger, J. Zhang and M. Steele. 2019. "Spatiotemporal variability of sea ice the Arctic's Last Ice Area." *Geophysical Research Letters* 46 (20): 11237–11243.

Mouginot, Jérémie, Eric Rignot, Anders A. Bjørk, Michiel van den Broeke, Romain Millan, Mathieu Morlighem, Brice Noël, Bernd Scheuchl and Michael Wood. 2019. "Forty-six years of Greenland Ice Sheet mass balance from 1972 to 2018." *PNAS* 116 (19): 9239–9244.

Nieuwenhuis, Martin and Aya Nassar. 2018. "Dust: Perfect circularity." *Cultural Geographies* 25 (3): 501–507.

Nordenskiöld, Alfred Erik. 1872. "Account of an expedition to Greenland in the year 1870." *Geological Magazine* 9 (97): 289–306.

Nuttall, Mark. 1998. *Protecting the Arctic: Indigenous peoples and cultural survival.* Amsterdam: Harwood Academic Publishers.

Nuttall, Mark. 2017. *Climate, Society and Subsurface Politics in Greenland: Under the Great Ice.* London and New York: Routledge.

Nuttall, Mark. 2019. "Icy, watery, liquescent: Sensing and feeling climate change on Northwest Greenland's Coast." *Journal of Northern Studies* 14 (2): 71–91.

Nuttall, Mark. 2022a. "'A scene of indescribable horror': Peril, terror and shipwreck in Melville Bay, Northwest Greenland." In Jörge Vögele, Luisa Rittershaus, Timo Heimerdinger and Christoph auf der Horst (eds.) *The Cruel Sea*, Köln: Böhlau, pp. 29–38.

Nuttall, Mark. 2022b. "Places of memory, anticipation, and agitation in Northwest Greenland." In Kenneth L. Pratt and Scott A. Heyes (eds.) *Memory and Landscape: Indigenous responses to a changing North.* Athabasca: Athabasca University Press, pp. 157–177.

Orban, Clara E. 1997. *The Culture of Fragments. Words and images in futurism and surrealism.* Amsterdam and Atlanta: Editions Rodopi.

Peeken, Ilka, Sebastian Primpke, Birte Beyer, Julia Gütermann, Christian Katlein, Thomas Krumpen, Melanie Bergmann, Laura Hehemann and Gunnar Gerdts. 2018. "Arctic sea ice is an important temporal sink and means for transport for microplastic." *Nature Communications* 9: 1505. DOI: 10.1038/s41467-018-03825-5.

Pendleton, Simon L., Gifford H. Miller, Nathaniel Lifton, Scott J. Lehman, John Southon, Sarah E. Crump and Robert S. Anderson. 2019. "Rapidly receding Arctic Canada glaciers revealing landscapes continuously ice-covered for more than 40,000 years." *Nature Communications* 10 (1): 445.

Polyakov, Igor V., et al. 2017. "Greater role for Atlantic inflows on sea ice loss in the Eurasian basin of the arctic ocean." *Science* 356 (6335): 285–291.

Ramutsindela, Maano, Sylvain Guyot, Sébastian Boillat, Frédéric Giraut and Patrick Bottazi. 2020. "The geopolitics of protected areas." *Geopolitics* 25 (1): 240–266.

Rasmussen, Knud. 1921. *Greenland by the Polar Sea.* New York: Frederick A. Stokes Company.

Ribeiro, Sofia, Audrey Limoges, Guillaume Massé, Kasper L. Johannsen, William Colgan, Kaarina Weckström, Rebecca Jackson, Eleanor Georgiadis, Naja Mikkelsen, Antoon Kuijpers, Jesper Olsen, Steffen M. Olsen, Martin Nissen, Astrid Strunk, Sebastian Wetterich, Jari Syväranta, Andrew C.G. Henderson, Helen Mackay, Sami Taipale, Erik Jeppesen, Nicolaj K. Larsen, Xavier Crosta, Jacques Giraudeau, Mark Nuttall, Bjarne Grønnow, Anders Mosbech and Thomas A. Davidson. 2021. "Vulnerability of the North Water ecosystem to climate change." *Nature Communications* 12, 4475. https://doi.org/10.1038/s41467-021-24742-0.

Ryan, Jonathan C., Alun Hubbard, Marek Stibal, Tristram D. Irvine-Fynn, Joseph Cook, Laurence C. Smith, Karen Cameron and Jason Box. 2018. "Dark Zone of the Greenland Ice Sheet controlled by distributed biologically active impurities." *Nature Communications* 9, 1065. https://doi.org/10.1038/s41467-018-03353-2.

Simonsen, Marius Folden, Giovanni Baccolo, Thomas Blunier, Alejandra Borunda, Barbara Delmote, Robert Frei, Steven Goldstein, Aslak Grinsted, Helle Astrid Kjær, Todd Sowers, Anders Svensson, Bo Vinther, Diana Vladimirova and Gisela Wincklerm Mai Winstrup and Paul Vallelonga. 2019. "East Greenland ice core dust record reveals timing of Greenland ice sheet advance and retreat." *Nature Communications* 10, 4494. https://doi.org/10.1038/s41467-019-12546-2.

Speer, Lisa, Regan Nelson, Robbert Casier, Maria Gavrilo, Cecilie von Quillfeldt, Jesse Cleary, Patrick Halpin and Patricia Hooper. 2017. *Natural Marine World Heritage in the Arctic Ocean*. Report of an expert workshop and review process. Gland: IUCN.

Stroeve, Julienne C., Mark C. Serreze, Marika M. Holland, Jennifer E. Kay, James Malanik and Andrew P. Barrett. 2012. "The Arctic's rapidly shrinking sea ice cover: A research synthesis." *Climatic Change* 110 (3–4): 1005–1027.

Struebig, Matthew J., Tigga Kingston, Akbar Zubaid, Adura Mohd-Adnan and Stephen J. Rossiter. 2008. "Conservation value of forest fragments to Palaeotropical bats." *Biological Conservation* 141 (8): 2112–2126.

Thomas, Sophie. 2003. "Assembling history: Fragments and ruins." *European Romantic Review* 14 (2): 177–186.

Tobo, Yukata, Kouji Adachi, P aul J. DeMott, Thomas C.J. Hill, Douglas S. Hamilton, Natalie M. Mahowald, Naoko Nagatsuka, Sho Ohata, Jun Uetake, Yukata Kondo and Makoto Koike. 2019. "Glacially sourced dust as a potentially significant source of ice nucleating particles." *Nature Geoscience* 12 (4): 253–258.

Trusel, Luke D., Sarah B. Das, Matthew B. Osman, Matthew J. Evans, Ben E. Smith, Xavier Fettweis, Joseph R. McConnell. Brice B.Y. Noël and Michiel R. van den Broeke. 2018. "Nonlinear rise in Greenland runoff in response to post-industrial Arctic warming." *Nature* 564, 104–108. https://doi.org/10.1038/s41586-018-0752-4.

Wadhams, Peter. 2016. *A Farewell to Ice. A report from the Arctic*. London: Penguin.

Walsh, John E., Florence Fetterer, J. Scott Stewart and William L. Chapman. 2017. "A database for depicting Arctic sea ice variations back to 1850." *Geographical Review* 107 (1): 89–107.

Wang, Muyin and James E. Overland. 2009. "A sea ice free summer arctic within thirty years?" *Geophysical Research Letters* 36 (7). https://doi.org/10.1029/2009GL037820.

Wientjes, I.G.M., R.S.W. van de Wal, G.-J., Reichart, A. Sluijs and J. Oerlemans. 2011. "Dust from the dark region in the western ablation zone of the Greenland ice sheet." *The Cryosphere* 5 (3): 589–601.

Wilkinson, Jeremy and Julienne Stroeve. 2018. "Polar sea ice as a barometer and driver of change." In Mark Nuttall, Torben R. Christensen and Martin J. Siegert (eds.) *The Routledge Handbook of the Polar Regions*, New York: Routledge, pp. 176–184.

Index

Note: *Italic* page numbers refer to figures.

For Product Safety Concerns and Information please contact our EU
representative GPSR@taylorandfrancis.com
Taylor & Francis Verlag GmbH, Kaufingerstraße 24, 80331 München, Germany

* 9 7 8 1 0 3 2 0 0 7 5 1 9 *